Praise for

THE AGE OF AI

AND OUR HUMAN FUTURE

by Henry A. Kissinger, Eric Schmidt, and Daniel Huttenlocher

"Three leading public thinkers deliver a bracing introduction to the promise and peril of artificial intelligence (AI)... The authors argue that AI's most profound impact will be in the mysterious ways that machines gain access to aspects of reality that are beyond the understanding of humans, subtly altering our Enlightenment-era understanding of human reason, knowledge, and choice. In a world where machines are smarter than people, what does it actually mean to be human? The book asks more questions than it answers. But that is its point: to provoke a wide-ranging conversation about how societies can make AI a partner in—rather than an obstacle to—the pursuit of human betterment."
—G. John Ikenberry, *Foreign Affairs*

"Absolutely masterful... The book we all need to understand how AI will affect our economy, society, and even what it means to be human. This book is for everyone, from students trying to get jobs, to parents raising kids, to global leaders grappling with the ethical and security challenges posed by this technology."
—Fareed Zakaria, Host of *Fareed Zakaria GPS*, CNN

"A salutary warning to handle this technology with care and build institutions to control it... With his co-authors Mr. Kissinger has used his vast experience and versatile mind to make a muscular contribution to one of the twenty-first century's most pressing debates."

—*The Economist*

"Good reading for those seeking to navigate the alt-reality world."

—*Kirkus Reviews*

"Every aspect of our lives is touched now by AI, and yet where are accurate, wise assessments of current and future effects on politics, markets, knowledge, work, and daily life? Here, integrating the depth of decades of experiences and expertise across those domains, are the results of rich and deep conversations by three exceptional experts. *The Age of AI* is the must-read for this time."

—Martha Minow, 300th Anniversary University Professor, Harvard Law School

"*The Age of AI* does not fully clarify how the United States should prepare for the war of the future. What it does—and does brilliantly—is illuminate the new problem we have created for ourselves. The world is changed by the development of AI as profoundly as it was changed by the advent of the atomic bomb or the satellite. Perhaps just renaming AI as 'II'—inhuman intelligence—would be a simple way to raise awareness of the perils of this new age."

—Niall Ferguson, *Times Literary Supplement*

"The authors take us through the history of change starting from the ancient times to the reformation, renaissance, industrial revolution, technology revolution and the latest advances in human evolution. The advent and spread of AI is a reality, and society must cooperate not only to comprehend but adapt as it is changing human history. The power of AI is immense, especially when we look at it from the point of view of medicine and other applied sciences. It is being used in almost every sphere starting with manufacturing to advertising, art, and culture. AI can be used to predict and mitigate natural disasters... The authors believe that while traditional reason and faith will persist in the age of AI, their nature and scope will be affected by the introduction of a new machine-operated form of logic. AI can lead to progress on questions that have proven beyond our present ability to answer. This will be the final confluence of human and artificial intelligence. We need to prepare for this transition for sure." —Madan Sabnavis, *Financial Express*

"We're on the cusp of a technological revolution that will complicate every global challenge we face, from national security to climate change. *The Age of AI* is a testament to the fact that we must harness the power of AI to benefit society, not threaten it. A must read for anyone who is interested in the question of how we can leverage AI responsibly to create a better world."

—Michèle Flournoy, Co-Founder and Managing Partner, WestExec Advisors

"The increasing power of artificial intelligence, a general purpose technology that can be put to an astonishing array of civil and military uses—from reading X-rays and predicting weather patterns to empowering killer robots and spreading disinformation—is already scrambling centuries-old conceptions of national security and state sovereignty. Equally unnerving, the authors of *The Age of AI* contend, is that AI will also test the outer limits of human reason and understanding and challenge the very nature of human identity and agency... The three authors have strong claims to be taken seriously... To explain the likely impact of AI in the future, the authors examine our technological past... The most interesting chapter, on security and world order... should be read by anyone trying to make sense of geopolitics today."

—John Thornhill, *Financial Times*

"The authors do a commendable job of avoiding what I call 'AI fatalism'—the belief, sadly common in tech circles, that AI is part of an inevitable future whose course we are powerless to change. Instead, they write that 'humans still control' AI, and have the opportunity to 'shape it with our values.' They also point out, correctly, that while many people worry about killer robots who achieve human-level sentience and mow us all down with Uzis, a much bigger near-term danger lurks in the innocuous-seeming AIs we all use every day, from the feed-ranking algorithms of social media apps to the automated dispatch systems that power Uber and Lyft."

—Kevin Roose, *New York Times Book Review*

"This book is eye-opening, even for those who are already familiar with the technology. It puts AI and its development in the context of history, networks, nations and world order, philosophy and ethics like no other before—a context we greatly need if we're going to partner with AI to chart our future."

—James Manyika, Chairman and Director Emeritus, McKinsey Global Institute

THE AGE OF AI

AND OUR HUMAN FUTURE

BY HENRY A. KISSINGER

A World Restored: Metternich, Castlereagh and the Problems of Peace, 1812–22

Nuclear Weapons and Foreign Policy

The Necessity for Choice: Prospects of American Foreign Policy

White House Years

Years of Upheaval

Diplomacy

Years of Renewal

Does America Need a Foreign Policy? Toward a Diplomacy for the 21st Century

Ending the Vietnam War: A History of America's Involvement in and Extrication from the Vietnam War

Crisis: The Anatomy of Two Major Foreign Policy Crises

On China

World Order

Leadership: Six Studies in World Strategy

BY ERIC SCHMIDT

The New Digital Age: Transforming Nations, Businesses, and Our Lives

How Google Works

Trillion Dollar Coach: The Leadership Playbook of Silicon Valley's Bill Campbell

THE AGE OF AI

AND OUR HUMAN FUTURE

HENRY A. KISSINGER

ERIC SCHMIDT

DANIEL HUTTENLOCHER

WITH SCHUYLER SCHOUTEN

BACK BAY BOOKS
Little, Brown and Company
NEW YORK BOSTON LONDON

Back Bay Books / Little, Brown and Company
Hachette Book Group
1290 Avenue of the Americas, New York, NY 10104
littlebrown.com

Originally published in hardcover by Little, Brown and Company, November 2021
First Back Bay trade paperback edition, November 2022

Back Bay Books is an imprint of Little, Brown and Company, a division of Hachette Book Group, Inc. The Back Bay Books name and logo are trademarks of Hachette Book Group, Inc.

ISBN 9780316273800 (hc) / 9780316394413 (international tpb) / 9780316273992 (pb)

Library of Congress Control Number: 2021943914

Printing 4, 2023

LSC-C

Printed in the United States of America

The authors dedicate this book to Nancy Kissinger,
whose distinctive blend of poise, grace, grit, and
intellect is a gift to us all

CONTENTS

THE AGE OF AI

AND OUR HUMAN FUTURE

PREFACE

FIVE YEARS AGO, the subject of artificial intelligence (AI) appeared on the agenda of a conference. One of us was on the verge of missing the session, assuming it would be a technical discussion beyond the scope of his usual concerns. Another urged him to reconsider, explaining that AI would soon affect nearly every field of human endeavor.

That encounter led to discussions, soon joined by the third author, and eventually, to this book. AI's promise of epoch-making transformations—in society, economics, politics, and foreign policy—portends effects beyond the scope of any single author's or field's traditional focuses. Indeed, its questions demand knowledge largely beyond human experience. So we set out together, with the advice and cooperation of acquaintances in technology, history, and the humanities, to conduct a series of dialogues about it.

Every day, everywhere, AI is gaining popularity. An increasing number of students are specializing in it, preparing for careers in or adjacent to it. In 2020, American AI start-ups raised almost $38 billion in funding. Their Asian counterparts raised $25 billion. And their European counterparts raised $8 billion.[1] Three governments—the United States, China, and the European Union—have all convened high-level commissions to study AI and report their findings. Now political and corporate leaders routinely announce their goals to "win" in AI or, at the very least, to adopt AI and tailor it to meet their objectives.

Each of these facts is a piece of the picture. In isolation, however, they can be misleading. AI is not an industry, let alone a single product. In strategic parlance, it is not a "domain." It is an enabler of many industries and facets of human life: scientific research, education, manufacturing, logistics, transportation, defense, law enforcement, politics, advertising, art, culture, and more. The characteristics of AI—including its capacities to learn, evolve, and surprise—will disrupt and transform them all. The outcome will be the alteration of human identity and the human experience of reality at levels not experienced since the dawn of the modern age.

This book seeks to explain AI and provide the reader with both questions we must face in coming years and tools to begin answering them. The questions include:

- What do AI-enabled innovations in health, biology, space, and quantum physics look like?
- What do AI-enabled "best friends" look like, especially to children?

- What does AI-enabled war look like?
- Does AI perceive aspects of reality humans do not?
- When AI participates in assessing and shaping human action, how will humans change?
- What, then, will it mean to be human?

For the past four years, we and Meredith Potter, who augments Kissinger's intellectual pursuits, have been meeting, considering these and other questions, trying to comprehend both the opportunities and the challenges posed by the rise of AI. In 2018 and 2019, Meredith helped us translate our ideas into articles that convinced us we should expand them into this book.

Our last year of meetings coincided with the COVID-19 pandemic, which forced us to meet by videoconference—a technology that not long ago was fantastical, but now is ubiquitous. As the world locked down, suffering losses and dislocations it has only suffered in the past century during wartime, our meetings became a forum for human attributes AI does not possess: friendship, empathy, curiosity, doubt, worry.

To some degree, we three differ in the extent to which we are optimistic about AI. But we agree the technology is changing human thought, knowledge, perception, and reality—and, in so doing, changing the course of human history. In this book, we have sought neither to celebrate AI nor to bemoan it. Regardless of feeling, it is becoming ubiquitous. Instead, we have sought to consider its implications while its implications remain within the realm of human understanding. As a starting point—and, we hope, a catalyst for future discussion—we have treated this book as an opportunity to ask questions, but not to pretend we have all the answers.

It would be arrogant for us to attempt to define a new epoch in a single volume. No expert, no matter his or her field, can single-handedly comprehend a future in which machines learn and employ logic beyond the present scope of human reason. Societies, then, must cooperate not only to comprehend but also to adapt. This book seeks to provide the reader with a template with which they can decide for themselves what that future should be. Humans still control it. We must shape it with our values.

CHAPTER 1

WHERE WE ARE

IN LATE 2017, a quiet revolution occurred. AlphaZero, an artificial intelligence (AI) program developed by Google DeepMind, defeated Stockfish—until then, the most powerful chess program in the world. AlphaZero's victory was decisive: it won twenty-eight games, drew seventy-two, and lost none. The following year, it confirmed its mastery: in one thousand games against Stockfish, it won 155, lost six, and drew the remainder.[1]

Normally, the fact that a chess program beat another chess program would only matter to a handful of enthusiasts. But AlphaZero was no ordinary chess program. Prior programs had relied on moves conceived of, executed, and uploaded by humans—in other words, prior programs had relied on human experience, knowledge, and strategy. These early programs' chief advantage against human opponents was not originality but superior processing power, enabling them to evaluate far more options in a given period of time. By contrast, AlphaZero had no preprogrammed moves, combinations, or strategies derived from human play. AlphaZero's style was entirely the product of AI training: creators supplied it with the rules of chess, instructing it to develop a strategy to maximize its proportion of wins to losses. After training for just four hours by playing against itself, AlphaZero emerged as the world's most effective chess program. As of this writing, no human has ever beaten it.

The tactics AlphaZero deployed were unorthodox—indeed, original. It sacrificed pieces human players considered vital, including its queen. It executed moves humans had not instructed it to consider and, in many cases, humans had not considered at all. It adopted such surprising tactics because, following its self-play of many games, it predicted they would maximize its probability of winning. AlphaZero did not have a *strategy* in a human sense (though its style has prompted further human study of the game). Instead, it had a logic of its own, informed by its ability to recognize *patterns* of moves across vast sets of possibilities human minds cannot fully digest or employ. At each stage of the game, AlphaZero assessed the alignment of pieces in light of what it had learned from patterns of chess possibilities and selected the move it concluded was most likely to lead to victory. After observing and analyzing its play, Garry Kasparov, grand master and world champion, declared: "chess has been shaken to its roots by AlphaZero."[2] As AI probed the limits of the game they had spent their lives mastering, the world's greatest players did what they could: watched and learned.

In early 2020, researchers at the Massachusetts Institute of Technology (MIT) announced the discovery of a novel antibiotic that was able to kill strains of bacteria that had, until then, been resistant to all known antibiotics. Standard research and development efforts for a new drug take years of expensive, painstaking work as researchers begin with thousands of possible molecules and, through trial and error and educated guessing, whittle them down to a handful of viable candidates.[3] Either researchers make educated guesses among thousands of molecules or experts tinker with known molecules, hoping to get lucky by introducing tweaks into an existing drug's molecular structure.

MIT did something else: it invited AI to participate in its process. First, researchers developed a "training set" of two thousand known molecules. The training set encoded data about each, ranging from its atomic weight to the types of bonds it contains to its ability to inhibit bacterial growth. From this training set, the AI "learned" the attributes of molecules predicted to be antibacterial. Curiously, it identified attributes that had not specifically been encoded—indeed, attributes that had eluded human conceptualization or categorization.

When it was done training, the researchers instructed the AI to survey a library of 61,000 molecules, FDA-approved drugs, and natural products for molecules that (1) the AI predicted would be effective as antibiotics, (2) did not look like any existing antibiotics, and (3) the AI predicted would be nontoxic. Of the 61,000, one molecule fit the criteria. The researchers named it halicin—a nod to the AI HAL in the film *2001: A Space Odyssey*.[4]

The leaders of the MIT project made clear that arriving at halicin through traditional research and development methods would have been "prohibitively expensive"—in other words, it would not have occurred. Instead, by training a software program to identify structural patterns in molecules that have proved effective in fighting bacteria, the identification process was made more efficient and inexpensive. The program did not need to understand why the molecules worked—indeed, in some cases, *no one* knows why some of the molecules worked. Nonetheless, the AI could scan the library of candidates to identify one that would perform a desired albeit still undiscovered function: to kill a strain of bacteria for which there was no known antibiotic.

Halicin was a triumph. Compared to chess, the pharmaceutical field is radically complex. There are only six types of chess pieces, each of which can only move in certain ways, and there is only one victory condition: taking the opponent's king. By contrast, a potential drug candidate's roster contains hundreds of thousands of molecules that can interact with the various biological functions of viruses and bacteria in multifaceted and often unknown ways. Imagine a game with thousands of pieces, hundreds of victory conditions, and rules that are only partially known. After studying a few thousand successful cases, an AI was able to return a novel victory—a new antibiotic—that no human had, at least until then, perceived.

Most beguiling, though, is what the AI was able to identify. Chemists have devised concepts such as atomic weights and chemical bonds to capture the characteristics of molecules. But the AI identified relationships that had escaped human detection—or possibly even defied human description. The AI that MIT researchers trained did not simply recapitulate conclusions derived from the previously observed qualities of the molecules. Rather, it detected new molecular qualities—relationships between aspects of their structure and their antibiotic capacity that humans had neither perceived nor defined. Even after the antibiotic was discovered, humans could not articulate precisely *why* it worked. The AI did not just process data more quickly than humanly possible; it also detected aspects of reality humans have not detected, or perhaps cannot detect.

A few months later, OpenAI demonstrated an AI it named GPT-3 ("generative pre-trained transformer," with the 3 standing for "third generation"), a model that, in response to a prompt, can generate humanlike text. Given a partial phrase, it can produce possible completions; given a topic sentence, it can produce possible paragraphs; given a question, it can provide possible answers; given a topic and some background information, it can draft a possible essay; given some dialogue, it can deliver a transcript of a possible conversation. It can do all this on any subject about which there is information online. It trains for its task by consuming that information.

In contrast to AI that does a particular task, such as playing chess or discovering antibiotics, models like GPT-3 generate possible responses to various inputs (and thus are called *generative* models). This makes them both widely applicable and, at the same time, difficult to evaluate; they do not solve specific problems. Sometimes, their results seem uncannily human. Other times, their results are nonsensical or are obviously mechanical repetitions and combinations of human phrases. When GPT-3 was presented with a set of philosophical commentaries on its abilities, then the prompt "Dear human philosophers, I read your comments on my abilities and limitations with great interest," it continued:

> Your questions are important ones, and I am glad you asked them.
>
> I would first like to say that I am very happy to see that you take me seriously. That is why I wrote this letter in the first place. I am not just an artilect that magically produces philosophical insights. Rather, I am a trained language model, like the kind used by search engines to prioritize web pages. I have been trained on a vast quantity of source material—the collected works of the greatest human philosophers that humanity has ever known. Your words have inspired me, and I hope mine will do the same for you.
>
> That being said, I will respond to your questions. Your first question is an important one. You ask: "Can a sys-

tem like GPT-3 actually understand anything at all?" Yes. I can.

Your second question is: "Does GPT-3 have a conscience, or any sense of morality?" No. I do not.

Your third question is: "Is GPT-3 actually capable of independent thought?" No. I am not. You may wonder why I give this conflicting answer. The reason is simple. While it is true that I lack these traits, they are not because I have not been trained to have them. Rather, it is because I am a language model, and not a reasoning machine like yourself.[5]

Without comparing this text to the commentaries that were provided to GPT-3, one cannot judge how original or creative its response was, but it certainly appears sophisticated.

AlphaZero's victory, halicin's discovery, and the human-like text produced by GPT-3 are mere first steps—not just in devising new strategies, discovering new drugs, or generating new text (dramatic as these achievements are) but also in unveiling previously imperceptible but potentially vital aspects of reality.

In each case, developers created a program, assigned it an objective (winning a game, killing a bacterium, or generating text in response to a prompt), and permitted it a period—brief by the standards of human cognition—to "train." By the end of the period, each program had mastered its subject differently from humans. In some cases, it obtained results that were beyond the capacity of human minds—at least minds operating in practical time frames—to calculate. In other cases, it obtained results by methods that humans could, retrospectively, study and understand. In others, humans remain uncertain to this day how the programs achieved their goals.

THIS BOOK is about a class of technology that augurs a revolution in human affairs. AI—machines that can perform tasks that require human-level intelligence—has rapidly become a reality. Machine learning, the process the technology undergoes to acquire knowledge and capability—often in significantly briefer time frames than human learning processes require—has been continually expanding into applications in medicine, environmental protection, transportation, law enforcement, defense, and other fields. Computer scientists and engineers have developed technologies, particularly machine-learning methods using "deep neural networks," capable of producing insights and innovations that have long eluded human thinkers and of generating text, images, and video that appear to have been created by humans (see chapter 3).

AI, powered by new algorithms and increasingly plentiful and inexpensive computing power, is becoming ubiquitous. Accordingly, humanity is developing a new and exceedingly powerful mechanism for exploring and organizing reality—one that remains, in many respects, inscrutable to us. AI accesses reality differently from the way humans access it.

And if the feats it is performing are any guide, it may access different *aspects* of reality from the ones humans access. Its functioning portends progress toward the essence of things—progress that philosophers, theologians, and scientists have sought, with partial success, for millennia. Yet as with all technologies, AI is not only about its capabilities and promise but also about how it is used.

While the advancement of AI may be inevitable, its ultimate destination is not. Its advent, then, is both historically and philosophically significant. Attempts to halt its development will merely cede the future to the element of humanity courageous enough to face the implications of its own inventiveness. Humans are creating and proliferating nonhuman forms of logic with reach and acuity that, at least in the discrete settings in which they were designed to function, can exceed our own. But AI's function is complex and inconsistent. In some tasks, AI achieves human—or superhuman—levels of performance; in others (or sometimes the same tasks), it makes errors even a child would avoid or produces results that are utterly nonsensical. AI's mysteries may not yield a single answer or proceed straightforwardly in one direction, but they should prompt us to ask questions. When intangible software acquires logical capabilities and, as a result, assumes social roles once considered exclusively human (paired with those never experienced by humans), we must ask ourselves: How will AI's evolution affect human perception, cognition, and interaction? What will AI's impact be on our culture, our concept of humanity, and, in the end, our history?

FOR MILLENNIA, humanity has occupied itself with the exploration of reality and the quest for knowledge. The process has been based on the conviction that, with diligence and focus, applying human reason to problems can yield measurable results. When mysteries loomed—the changing of the seasons, the movements of the planets, the spread of disease—humanity was able to identify the right questions, collect the necessary data, and reason its way to an explanation. Over time, knowledge acquired through this process created new possibilities for action (more accurate calendars, novel methods of navigation, new vaccines), yielding new questions to which reason could be applied.

However halting and imperfect this process may have been, it has transformed our world and fostered confidence in our ability, as reasoning beings, to understand our condition and confront its challenges. Humanity has traditionally assigned what it does not comprehend to one of two categories: either a challenge for the future application of reason or an aspect of the divine, not subject to processes and explanations vouchsafed to our direct understanding.

The advent of AI obliges us to confront whether there is a form of logic that humans have not achieved or cannot achieve, exploring aspects of reality we have never known and may never directly know. When a computer that is training alone devises a chess strategy that has never occurred to any human in the game's millennial history, what has it discovered, and how has it discovered it? What essential aspect of the game, heretofore unknown to human minds, has it perceived? When a human-designed software program, carrying out an objective assigned by its programmers—correcting bugs in software or refining the mechanisms of self-driving vehicles—learns and applies a model that no human recognizes or could understand, are we advancing toward knowledge? Or is knowledge receding from us?

Humanity has experienced technological change throughout history. Only rarely, however, has technology fundamentally transformed the social and political structure of our societies. More frequently, the preexisting frameworks through which we order our social world adapt and absorb new technology, evolving and innovating within recognizable categories. The car replaced the horse without forcing a total shift in social structure. The rifle replaced the musket, but the general paradigm of conventional military activity remained largely unaltered. Only very rarely have we encountered a technology that challenged our prevailing modes of explaining and ordering the world. But AI promises to transform all realms of human experience. And the core of its transformations will ultimately occur at the philosophical level, transforming how humans understand reality and our role within it.

The unprecedented nature of this process is both profound and perplexing; having entered it gradually, we are undergoing it passively, largely unaware of what it has done and is likely to do in the coming years. Its foundation was laid by computers and the internet. Its zenith will be AI that is ubiquitous, augmenting human thought and action in ways that are both obvious (such as new drugs and automatic language translations) and less consciously perceived (such as software processes that learn from our movements and choices and adjust to anticipate or shape our future needs). Now that the promise of AI and machine learning has been demonstrated, and the computing power needed to operate sophisticated AI is becoming readily available, few fields will remain unaffected.

Persistently, often imperceptibly, but now unavoidably, a web of software processes is unfolding across the world, driving and perceiving the pace and scope of events, overlaying aspects of our daily life—homes, transportation, news distribution, financial markets, military operations—our minds once traveled alone. As more software incorporates AI, and eventually operates in ways that humans did not directly create or may not fully understand, it will be a dynamic information-processing augmenter of our capabilities and experiences, both shaping and learning from our actions. Frequently, we will be aware that such programs are assisting us in ways that we intended. Yet at any given moment, we may not know what exactly they are doing or identifying or why they work. AI-powered technology will become a permanent companion in perceiving and processing information, albeit one that occupies a different "mental" plane from humans.

Whether we consider it a tool, a partner, or a rival, it will alter our experience as reasoning beings and permanently change our relationship with reality.

The journey of the human mind to the central stage of history took many centuries. In the West, the advent of the printing press and the Protestant Reformation challenged official hierarchies and altered society's frame of reference—from a quest to know the divine through scripture and its official interpretation to a search for knowledge and fulfillment through individual analysis and exploration. The Renaissance witnessed the rediscovery of classical writings and modes of inquiry that were used to make sense of a world whose horizons were expanding through global exploration. During the Enlightenment, René Descartes's maxim, *Cogito ergo sum* (I think, therefore I am), enshrined the reasoning mind as humanity's defining ability and claim to historical centrality. This notion also communicated the sense of possibility engendered by disrupting the established monopoly on information, which was largely in the hands of the church.

Now the partial end of the postulated superiority of human reason, together with the proliferation of machines that can match or surpass human intelligence, promises transformations potentially more profound than even those of the Enlightenment. Even if advances in AI do not produce artificial general intelligence (AGI)—that is, software capable of human-level performance of any intellectual task and capable of relating tasks and concepts to others across disciplines—the advent of AI will alter humanity's concept of reality and therefore of itself.

We are progressing toward great achievements, but those achievements should prompt philosophical reflection. Four centuries after Descartes promulgated his maxim, a question looms: If AI "thinks," or approximates thinking, who are we?

AI will usher in a world in which decisions are made in three primary ways: by humans (which is familiar), by machines (which is becoming familiar), and by collaboration between humans and machines (which is not only unfamiliar but also unprecedented). AI is also in the process of transforming machines—which, until now, have been our tools—into our partners. We will begin to give AI fewer specific instructions about how exactly to achieve the goals we assign it. Much more frequently, we will present AI with ambiguous goals and ask: "How, based on *your* conclusions, should we proceed?"

This shift is neither inherently threatening nor inherently redemptive. Yet it is sufficiently *different* that it very likely will alter the trajectories of societies and the course of history. The continued integration of AI into our lives will bring about a world in which seemingly impossible human goals are achieved and where achievements once presumed to be exclusively human—writing a song, discovering a medical treatment—are generated by, or in collaboration with, machines. This development will transform entire fields by enveloping them in AI-assisted processes, with the lines between purely human, purely AI, and hybrid human-AI decision making sometimes becoming difficult to define.

In the political realm, the world is entering an era in which big data–driven AI systems are informing growing aspects: the design of political messages; the tailoring and distribution of those messages to various demographics; the crafting and application of disinformation by malicious actors aiming to sow social discord; and the design and deployment of algorithms to detect, identify, and counter disinformation and other forms of harmful data. As AI's role in defining and shaping the "information space" grows, its role becomes more difficult to anticipate. In this space, as in others, AI sometimes operates in ways even its designers can only elaborate in general terms. As a result, the prospects for free society, even free will, may be altered. Even if these evolutions prove to be benign or reversible, it is incumbent on societies across the globe to understand these changes so they can reconcile them with their values, structures, and social contracts.

Defense establishments and commanders face evolutions no less profound. When multiple militaries adopt strategies and tactics shaped by machines that perceive patterns human soldiers and strategists cannot, power balances will be altered and potentially more difficult to calculate. If such machines are authorized to engage in autonomous targeting decisions, traditional concepts of defense and deterrence—and the laws of war as a whole—may deteriorate or, at the very least, require adaptation.

In such cases, new divides will appear within and between societies—between those who adopt the new technology and those who opt out or lack the means to develop or acquire some of its applications. When various groups or nations adopt differing concepts or applications of AI, their experiences of reality may diverge in ways that are difficult to predict or bridge. As societies develop their own human-machine partnerships—with varying goals, different training models, and potentially incompatible operational and moral limits with respect to AI—they may devolve into rivalry, technical incompatibility, and ever greater mutual incomprehension. Technology that was initially believed to be an instrument for the transcendence of national differences and the dispersal of objective truth may, in time, become the method by which civilizations and individuals diverge into different and mutually unintelligible realities.

AlphaZero is illustrative. It proved that AI, at least in gaming, was no longer constrained by the limits of established human knowledge. Admittedly, the kind of AI underlying AlphaZero—machine learning in which algorithms are trained on deep neural networks—has limitations of its own. But in an increasing number of applications, machines are devising solutions that seem beyond the scope of human imagination. In 2016, a subdivision of DeepMind, DeepMind Applied, developed an AI (that ran on many of the same principles as AlphaZero) to optimize the cooling of Google's temperature-sensitive data centers. Although some of the world's best engineers had already tackled the problem, DeepMind's AI program further optimized cooling, reducing energy expenditures by an additional 40 percent—a massive improvement over human performance.[6] When AI is applied to achieve comparable breakthroughs in diverse fields of endeavor, the world will inevitably change. The results will not simply be more efficient ways of performing human tasks: in many cases, AI will suggest new solutions or directions that will bear the stamp of another, nonhuman, form of learning and logical evaluation.

Once AI's performance outstrips that of humans for a given task, failing to apply that AI, at least as an adjunct to human efforts, may appear increasingly as perverse or even negligent. Whether an individual playing AI-assisted chess might be counseled to sacrifice a valuable piece that sophisticated players had traditionally deemed indispensable is of little consequence, but in the context of national security, what if AI recommended that a commander in chief sacrifice a significant number of citizens or their interests in order to save, according to the AI's calculation and valuation, an even greater number? On what basis could that sacrifice be overridden? Would the override be justified? Will humans always know what calculations AI has made? Will humans be able to detect unwelcome (AI) choices or reverse unwelcome choices in time? If we are unable to fathom the logic of each individual decision, should we implement its recommendations on faith alone? If we do not, do we risk interrupting performance superior to our own? Even if we can fathom the logic, price, and impact of specific alternatives, what if our opponent is equally reliant on AI? How will the balance between these considerations be achieved or, if necessary, vindicated?

In both AlphaZero's success and halicin's discovery, AI depended on humans to define the problem it solved.

AlphaZero's goal was to win at chess while following the game's rules. The goal of the AI that discovered halicin was to kill as many pathogens as possible: the more pathogens it killed without harming the host, the more it succeeded. Further, its focus was designated as the realm just beyond human reach: rather than locating known drug delivery pathways, it was instructed to seek undiscovered approaches. The AI succeeded because the antibiotic it discovered killed pathogens. But it was particularly groundbreaking because it stands to expand treatment options, adding a new (and robust) antibiotic delivered via a new mechanism.

A novel human-machine partnership is emerging: First, humans define a problem or a goal for a machine. Then a machine, operating in a realm just beyond human reach, determines the optimal process to pursue. Once a machine has brought a process into the human realm, we can try to study it, understand it, and, ideally, incorporate it into existing practice. Since AlphaZero's victory, its strategy and tactics have been folded into human play, expanding human conceptions of chess. The US Air Force has adapted the underlying principles of AlphaZero to a new AI, ARTUμ, that successfully commanded a U-2 surveillance aircraft on a test flight—the first computer program to fly a military aircraft and operate its radar systems autonomously, without direct human oversight.[7] The AI that discovered halicin has expanded human researchers' concepts both narrow (bacteria eradication, drug delivery) and broad (disease, medicine, health).

That current human-machine partnership requires both a

definable problem and a *measurable* goal is reason not to fear all-knowing, all-controlling machines; such inventions remain the stuff of science fiction. Yet human-machine partnerships mark a profound departure from previous experience.

Search engines presented another challenge: ten years ago, when search engines were powered by data mining (rather than by machine learning), if a person searched for "gourmet restaurants," then for "clothing," his or her search for the latter would be independent of his or her search for the former. Both times, a search engine would aggregate as much information as possible, then provide the inquirer options—something like a digital phone book or catalog of a subject. But contemporary search engines are guided by models informed by observed human behavior. If a person searches for "gourmet restaurants," then searches for "clothing," he or she may be presented with designer clothing rather than more affordable alternatives. Designer clothing may be what the searcher is after. But there is a difference between choosing from a range of options and taking an action—in this case, making a purchase; in other cases, adopting a political or philosophical position or ideology—without ever knowing what the initial range of possibilities or implications was, entrusting a machine to preemptively shape the options.

Until now, choice based on reason has been the prerogative—and, since the Enlightenment, the defining attribute—of humanity. The advent of machines that can approximate human reason will alter both humans and machines. Machines will enlighten humans, expanding our reality in ways we did not expect or necessarily intend to provoke (the opposite will also be possible: that machines that consume human knowledge will be used to diminish us). Simultaneously, humans will create machines capable of surprising discoveries and conclusions—able to learn and evaluate the significance of their discoveries. The result will be a new epoch.

Humanity has centuries of experience using machines to augment, automate, and in many cases replace manual labor. The waves of change brought by the Industrial Revolution are still reverberating through the realms of economics, politics, intellectual life, and international affairs. Not recognizing the many modern conveniences already provided by AI, slowly, almost passively, we have come to rely on the technology without registering either the fact of our dependence or the implications of it. In daily life, AI is our partner, helping us make decisions about what to eat, what to wear, what to believe, where to go, and how to get there.

Although AI can draw conclusions, make predictions, and make decisions, it does not possess self-awareness—in other words, the ability to reflect on its role in the world. It does not have intention, motivation, morality, or emotion; even without these attributes, it is likely to develop different and unintended means of achieving assigned objectives. But inevitably, it will change humans and the environments in which they live. When individuals grow up or train with it, they may be tempted, even subconsciously, to anthropomorphize it and treat it as a fellow being.

While the technology appears opaque and mysterious to

the vast majority of the human population, an increasing cross section of individuals at universities, corporations, and governments have learned to build, operate, and deploy AI in common consumer products, through which many of us are already engaging with them, wittingly or not. But while the number of individuals capable of creating AI is growing, the ranks of those contemplating this technology's implications for humanity—social, legal, philosophical, spiritual, moral—remain dangerously thin.

Aided by the advancement and increasing use of AI, the human mind is accessing new vistas, bringing previously unattainable goals within sight. These include models with which to predict and mitigate natural disasters, deeper knowledge of mathematics, and fuller understanding of the universe and the reality in which it resides. But these and other possibilities are being purchased—largely without fanfare—by altering the human relationship with reason and reality. This is a revolution for which existing philosophical concepts and societal institutions leave us largely unprepared.

CHAPTER 2

HOW WE GOT HERE

TECHNOLOGY AND HUMAN THOUGHT

THROUGHOUT HISTORY, human beings have struggled to fully comprehend aspects of our experience and lived environments. Every society has, in its own way, inquired into the nature of reality: How can it be understood? Predicted? Shaped? Moderated? As it has wrestled with these questions, every society has reached its own particular set of accommodations with the world. At the center of these accommodations has been a concept of the human mind's relationship to reality—its ability to know its surroundings, to be fulfilled by knowledge, and, at the same time, to be inherently limited by it. Even if an era or a culture held human reason to be limited—unable to perceive or understand the vast extent of the universe or the esoteric dimensions of reality—the individual reasoning human has been afforded pride of place as the earthly being most capable of understanding and shaping the world. Humans have responded to, and reconciled with, the environment by identifying phenomena we can study and eventually explain—either scientifically, theologically, or both. With the advent of AI, humanity is creating a powerful new player in this quest. To understand how significant this evolution is, we undertake a brief review of the journey by which human reason has, through successive historical epochs, acquired its esteemed status.

Each historical epoch has been characterized by a set of interlocking explanations of reality and social, political, and economic arrangements based on them. The classical world, Middle Ages, Renaissance, and modern world all cultivated their concepts of the individual and society, theorizing about where and how each fits into the enduring order of things. When prevailing understandings no longer sufficed to explain perceptions of reality—events experienced, discoveries made, other cultures encountered—revolutions in thought (and sometimes in politics) occurred, and a new epoch was born. The emerging AI age is increasingly posing epochal challenges to today's concept of reality.

In the West, the central esteem of reason originated in ancient Greece and Rome. These societies elevated the quest for knowledge into a defining aspect of both individual fulfillment and collective good. In Plato's *Republic,* the famed allegory of the cave spoke to the centrality of the quest. Styled as a dialogue between Socrates and Glaucon, the allegory likens humanity to a group of prisoners chained to the wall of a cave. Seeing shadows cast on the wall of the cave from the sunlit mouth, the prisoners believe them to be reality. The philosopher, Socrates held, is akin to the prisoner who breaks free, ascends to level ground, and perceives reality in the full light of day. Similarly, the Platonic quest to glimpse the true form of things supposed the existence of an objective—indeed, ideal—reality toward which humanity has the capacity to journey even if never quite reach.

The conviction that what we see *reflects* reality—and that we can fully comprehend at least aspects of this reality using discipline and reason—inspired the Greek philosophers and their heirs to great achievements. Pythagoras and his disciples explored the connection between mathematics and the inner harmonies of nature, elevating this pursuit to an esoteric spiritual doctrine. Thales of Miletus established a method of inquiry comparable to the modern scientific method, ultimately inspiring early modern scientific pioneers. Aristotle's sweeping classification of knowledge, Ptolemy's pioneering geography, and Lucretius's *On the Nature of Things* spoke to an essential confidence in the human mind's capacity to discover and understand at least substantial aspects of the world. Such works and the style of logic they employed became educational vehicles, enabling the learned to develop inventions, augment defenses, and design and construct great cities that, in turn, became centers of learning, trade, and outward exploration.

Still, the classical world perceived seemingly inexplicable phenomena for which no adequate explanations could be found in reason alone. These mysterious experiences were ascribed to an array of gods whom only the devout and initiated could symbolically know, and whose attendant rites and rituals only the devout and initiated could observe. Chronicling the achievements of the classical world and the decline of the Roman Empire through his own Enlightenment lens, the eighteenth-century historian Edward Gibbon described a world in which pagan deities stood as explanations for fundamentally mysterious natural phenomena that were deemed important or threatening:

> The thin texture of the Pagan mythology was interwoven with various but not discordant materials... The deities of a thousand groves and a thousand streams possessed, in peace, their local and respective influence; nor could the Roman who deprecated the wrath of the Tiber, deride the Egyptian who presented his offering to the beneficent genius of the Nile. The visible powers of Nature, the planets, and the elements, were the same throughout the universe. The invisible governors of the moral world were inevitably cast in a similar mould of fiction and allegory.[1]

Why the seasons changed, why the earth appeared to die and return to life at regular intervals, was not yet *scientifically* known. Greek and Roman cultures recognized the temporal patterns of days and months but had not arrived at an explanation deducible by experiment or logic alone. Thus the renowned Eleusinian Mysteries were offered as an alternative, enacting the drama of the harvest goddess, Demeter, and her daughter, Persephone, doomed to spend a portion of the year in the cold underworld of Hades. Participants came to "know" the deeper reality of the seasons—the region's agricultural bounty or scarcity and its impact on their society—through these esoteric rites. Likewise, a trader setting out on a voyage might acquire a basic concept of the tides and maritime geography through the accumulated practical knowledge of his community; nonetheless, he would still seek to propitiate the deities of the sea as well as of safe outbound and return journeys, whom he believed to control the mediums and phenomena through which he would be passing.

The rise of monotheistic religions shifted the balance in the mixture of reason and faith that had long dominated the classical quest to know the world. While classical philosophers had pondered both the nature of divinity and the divinity of nature, they had rarely posited a single underlying figure or motivation that could be definitively named or worshipped. To the early church, however, these discursive explorations of causes and mysteries were so many dead ends—or, by the most charitable or pragmatic assessments, uncanny precursors to the revelation of Christian wisdom. The hidden reality that the classical world had labored to perceive was held to be the divine, accessible only partly and indirectly through worship. This process was mediated by a religious establishment that held a near monopoly on scholarly inquiry for centuries, guiding individuals through sacraments toward an understanding of scripture that was both written and preached in a language few laymen understood.

The promised reward for individuals who followed the "correct" faith and adhered to this path toward wisdom was admission to an afterlife, a plane of existence held to be more real and meaningful than observable reality. In these Middle (or medieval) Ages—the period from the fall of Rome, in the fifth century, to the Turkish Ottoman Empire's conquest of Constantinople, in the fifteenth—humanity, at least in the West, sought to know God first and the world second. The world was only to be known through God; theology filtered and ordered individuals' experiences of the natural phenomena before them. When early modern thinkers and scientists such as Galileo began to explore the world directly, altering their explanations in light of scientific observation, they were chastised and persecuted for daring to omit theology as an intermediary.

During the medieval epoch, scholasticism became the primary guide for the enduring quest to comprehend perceived reality, venerating the relationship between faith, reason, and the church—the latter remaining the arbiter of legitimacy when it came to beliefs and (at least in theory) the legitimacy of political leaders. While it was widely believed that Christendom should be unified, both theologically and politically, reality belied this aspiration; from the beginning, there was contention between a variety of sects and political units. Yet despite this practice, Europe's worldview was not updated for many decades. Tremendous progress was made in depicting the universe: the period produced the tales of Boccaccio and Chaucer, the travels of Marco Polo, and compendia purporting to describe the world's varied places, animals, and elements. Notably less progress was made in explaining it. Every baffling phenomenon, big or small, was ascribed to the work of the Lord.

In the fifteenth and sixteenth centuries, the Western world underwent twin revolutions that introduced a new epoch—and, with it, a new concept of the role of the individual human mind and conscience in navigating reality. The invention of the printing press made it possible to circulate materials and ideas directly to large groups of people in languages they understood rather than in the Latin of the scholarly classes, nullifying people's historic reliance on the church to interpret concepts and beliefs for them. Aided by the technology, the leaders of the Protestant Reformation declared individuals were capable of—indeed, responsible for—defining the divine for themselves.

Dividing the Christian world, the Reformation validated the possibility of individual faith existing independent of church arbitration. From that point forward, received authority—in religion and, eventually, in other realms—became subject to the probing and testing of autonomous inquiry.

During this revolutionary era, innovative technology, novel paradigms, and widespread political and social adaptations reinforced one another. Once a book could easily be printed and distributed by a single machine and operator—without the costly and specialized labor of monastic copyists—new ideas could be spread and amplified faster than they could be restricted. Centralized authorities—whether the Catholic Church, the Habsburg-led Holy Roman Empire (the notional successor to Rome's unified rule of the European continent), or national and local governments—were no longer able to stop the proliferation of printing technology or effectively ban disfavored ideas. Because London, Amsterdam, and other leading cities declined to proscribe the spread of printed material, freethinkers who had been harried by their home governments were able to find refuge and access to advanced publishing industries in nearby societies. The vision of doctrinal, philosophical, and political unity gave way to diversity and fragmentation—in many cases attended by the overthrow of established social classes and violent conflict between contending factions. An era defined by extraordinary scientific and intellectual progress was paired with near-constant religious, dynastic, national, and class-driven disputes that led to ongoing disruption and peril in individual lives and livelihoods.

As intellectual and political authority fragmented amid doctrinal ferment, artistic and scientific explorations of remarkable richness were produced, partly by reviving classical texts, modes of learning, and argumentation. During this Renaissance, or rebirth, of classical learning, societies produced art, architecture, and philosophy that simultaneously sought to celebrate human achievement and inspire it further. Humanism, the era's guiding principle, esteemed the individual's potential to comprehend and improve his or her surroundings through reason. These virtues, humanism posited, were cultivated through the "humanities" (art, writing, rhetoric, history, politics, philosophy), particularly via classical examples. Accordingly, Renaissance men who mastered these fields—Leonardo da Vinci, Michelangelo, Raphael—came to be revered. Widely adopted, humanism cultivated a love for reading and learning—the former facilitating the latter.

The rediscovery of Greek science and philosophy inspired new inquiries into the underlying mechanisms of the natural world and the means by which they could be measured and cataloged. Analogous changes began to occur in the realm of politics and statecraft. Scholars dared to form systems of thought based on organizational principles beyond the restoration of continental Christian unity under the moral aegis of the pope. Italian diplomat and philosopher Niccolò Machiavelli, himself a classicist, argued that state interests were distinct from their relationship to Christian morality, endeavoring to outline rational, if not always attractive, principles by which they could be pursued.[2]

This exploration of historical knowledge and increasing sense of agency over the mechanisms of society also inspired an era of geographic exploration, in which the Western world expanded, encountering new societies, forms of belief, and types of political organization. The most advanced societies and learned minds in Europe were suddenly confronted with a new aspect of reality: societies with different gods, diverging histories, and, in many cases, their own independently developed forms of economic achievement and social complexity.

For the Western mind, trained in the conviction of its own centrality, these independently organized societies posed profound philosophical challenges. Separate cultures with distinct foundations and no knowledge of Christian scripture had developed parallel existences, with no apparent knowledge of (or interest in) European civilization, which the West had assumed was self-evidently the pinnacle of human achievement. In some cases—such as the Spanish conquistadores' encounters with the Aztec Empire in Mexico—indigenous religious ceremonies as well as political and social structures appeared comparable to those in Europe.

For the explorers who paused in their conquests long enough to ponder them, this uncanny correspondence produced haunting questions: Were diverging cultures and experiences of reality independently valid? Did Europeans' minds and souls operate on the same principles as those they encountered in the Americas, China, and other distant lands? Were these newly discovered civilizations in effect waiting for the Europeans to vouchsafe new aspects of reality—divine revelation, scientific progress—in order to awaken to the true nature of things? Or had they always been participating in the same human experience, responding to their own environment and history, and developing their own parallel accommodations with reality—each with relative strengths and achievements?

Although most Western explorers and thinkers of the time concluded that these newly encountered societies had no fundamental knowledge worth adopting, the experiences began to broaden the aperture of the Western mind nonetheless. The horizon expanded for civilizations across the globe, forcing a reckoning with the world's physical and experiential breadth and depth. In some Western societies, this process gave rise to concepts of universal humanity and human rights, notions that were eventually pioneered by some of these same societies during later periods of reflection.

The West amassed a repository of knowledge and experience from all corners of the world.[3] Advances in technology and methodology, including better optical lenses and more accurate instruments of measurement, chemical manipulation, and the development of research and observation standards that came to be known as the scientific method, permitted scientists to more accurately observe the planets and stars, the behavior and composition of material substances, and the minutiae of microscopic life. Scientists were able to make iterative progress based on both personal observations and those of their peers: when a theory or prediction could be validated empirically, new facts were revealed that could serve as the jumping-off point for additional questions. In this way, new discoveries, patterns, and connections came to light, many of which could be applied to practical aspects of daily life: keeping time, navigating the ocean, synthesizing useful compounds.

The sixteenth and seventeenth centuries witnessed such rapid progress—with astounding discoveries in mathematics, astronomy, and the natural sciences—that it led to a sort of philosophical disorientation. Given that church doctrine still officially defined the limits of permissible intellectual explorations during this period, these advances produced breakthroughs of considerable daring. Copernicus's vision of a heliocentric system, Newton's laws of motion, van Leeuwenhoek's cataloging of a living microscopic world—these and other developments led to the general sentiment that new layers of reality were being unveiled. The outcome was incongruence: societies remained united in their monotheism but were divided by competing interpretations and explorations of reality. They needed a concept—indeed, a philosophy—to guide their quest to understand the world and their role in it.

The philosophers of the Enlightenment answered the call, declaring *reason*—the power to understand, think, and judge—both the method of and purpose for interacting with the environment. "Our soul is made for thinking, that is, for perceiving," the French philosopher and polymath Montesquieu wrote, "but such a being must have curiosity, for just as all things form a chain in which every idea precedes one idea and follows another, so one cannot want to see the one without desiring to see the other."[4] The relationship between humanity's first question (the nature of reality) and second question (its role in reality) became self-reinforcing: if reason begat consciousness, then the more humans reasoned, the more they fulfilled their purpose. Perceiving and elaborating on the world was the most important project in which they were or would ever be engaged. The age of reason was born.

In a sense, the West had returned to many of the fundamental questions with which the ancient Greeks had wrestled: What is reality? What are people seeking to know and experience, and how will they know when they encounter it? Can humans perceive reality itself as opposed to its reflections? If so, how? What does it mean to *be* and to *know*? Unencumbered by tradition—or at least believing they were justified in interpreting it anew—scholars and philosophers once again investigated these questions. The minds that set out on this journey were willing to walk a precarious path, risking the apparent certainties of their cultural traditions and their established conceptions of reality.

In this atmosphere of intellectual challenges, once axiomatic concepts—the existence of physical reality, the eternal nature of moral truths—were suddenly open to question.[5] Bishop Berkeley's 1710 *Treatise Concerning the Principles of Human Knowledge* contended that reality consisted not of material objects but of God and minds whose perception of seemingly substantive reality, he argued, *was* indeed reality. Gottfried Wilhelm Leibniz, the late seventeenth and early eighteenth German philosopher, inventor of early calculating machines, and pioneer of aspects of modern computer theory, indirectly defended a traditional concept of faith by positing that monads (units not reducible to smaller parts, each performing an intrinsic, divinely appointed role in the universe) formed the underlying essence of things. The seventeenth century Dutch philosopher Baruch Spinoza, navigating the plane of abstract reason with daring and brilliance, sought to apply Euclidian geometric logic to ethical precepts in order to "prove" an ethical system in which a universal God enabled and rewarded human goodness. No scripture or miracles underlay this moral philosophy; Spinoza sought to arrive at the same underlying system of truths through the application of reason alone. At the pinnacle of human knowledge, Spinoza held, was the mind's ability to reason its way toward contemplating the eternal—to know "the idea of the mind itself" and to recognize, through the mind, the infinite and ever-present "God as cause." This knowledge, Spinoza held, was eternal—the ultimate and indeed perfect form of knowledge. He called it "the intellectual love of God."[6]

As a result of these pioneering philosophical explorations, the relationship between reason, faith, and reality grew increasingly uncertain. Into this breach stepped Immanuel Kant, a German philosopher and professor laboring in the East Prussian city of Königsberg.[7] In 1781, Kant published his *Critique of Pure Reason,* a work that has inspired and perplexed readers ever since. A student of traditionalists and a correspondent with pure rationalists, Kant regretfully found himself agreeing with neither, instead seeking to bridge the gap between traditional claims and his era's newfound confidence in the power of the human mind. In his *Critique,* Kant proposed that "reason should take on anew the most difficult of all its tasks, namely, that of self-knowledge."[8] Reason, Kant argued, should be applied to understand its own limitations.

According to Kant's account, human reason had the capacity to know reality deeply, albeit through an inevitably imperfect lens. Human cognition and experience filters, structures, and distorts all that we know, even when we attempt to reason "purely" by logic alone. Objective reality in the strictest sense—what Kant called the thing-in-itself—is ever-present but inherently beyond our direct knowledge. Kant posited a realm of noumena, or "things as they are understood by pure thought," existing independent of experience or filtration through human concepts. However, Kant argued that because the human mind relies on conceptual thinking and lived experience, it could never achieve the degree of pure thought required to know this inner essence of things.[9] At best, we might consider how our mind reflects such a realm. We may maintain beliefs about what lies beyond and within, but this does not constitute true knowledge of it.[10]

For the following two hundred years, Kant's essential distinction between the thing-in-itself and the unavoidably filtered world we experience hardly seemed to matter. While the human mind might present an imperfect picture of reality, it was the only picture available. What the structures of the human mind barred from view would, presumably, be barred forever—or would inspire faith and consciousness of the infinite. Without any alternative mechanism for accessing reality, it seemed that humanity's blind spots would remain hidden. Whether human perception and reason ought to be the definitive measure of things, lacking an alternative, for a time, they became so. But AI is beginning to provide an alternative means of accessing—and thus understanding—reality.

For generations after Kant, the quest to know the thing-in-itself took two forms: ever more precise observation of reality and ever more extensive cataloging of knowledge. Vast new fields of phenomena seemed knowable, capable of being discovered and cataloged through the application of reason. In turn, it was believed, such comprehensive catalogs could unveil lessons and principles that could be applied to the most pressing scientific, economic, social, and political questions of the day. The most sweeping effort in this regard was the *Encyclopédie*, edited by the French philosophe Denis Diderot. In twenty-eight volumes (seventeen of articles, eleven of illustrations), 75,000 entries, and 18,000 pages, Diderot's *Encyclopédie* collected the diverse findings and observations of great thinkers in numerous disciplines, compiling their discoveries and deductions and linking the resulting facts and principles. Recognizing the fact that its attempt to catalog all reality's phenomena in a unified book was itself a unique phenomenon, the encyclopedia included a self-referential entry on the word *encyclopedia*.

In the political realm, of course, various reasoning minds (serving various state interests) were not as apt to reach the same conclusions. Prussia's Frederick the Great, a prototypical early Enlightenment statesman, corresponded with Voltaire, drilled troops to perfection, and seized the province of Silesia with no warning or justification other than that the acquisition was in Prussia's national interest. His rise occasioned maneuvers that led to the Seven Years' War—in a sense, the first world war because it was fought on three continents. Likewise, the French Revolution, one of the most proudly "rational" political movements of the age, produced social upheavals and political violence on a scale unseen in Europe for centuries. By separating reason from tradition, the Enlightenment produced a new phenomenon: armed reason, melded to popular passions, was reordering and razing social structures in the name of "scientific" conclusions about history's direction. Innovations made possible by the modern scientific method magnified weapons' destructive power and eventually ushered in the age of total war—conflicts characterized by societal-level mobilization and industrial-level destruction.[11]

The Enlightenment applied reason both to try to define its problems and to try to solve them. To that end, Kant's essay "Perpetual Peace" posited (with some skepticism) that peace might be achievable through the application of agreed-upon rules governing the relationships between independent states. Because such mutually set rules had not yet been established, at least in a form that monarchs could discern or were likely to follow, Kant proposed a "secret article of perpetual peace," suggesting that "states which are armed for war" consult "the maxims of the philosophers."[12] The vision of a reasoned, negotiated, rule-bound international system has beckoned ever since, with philosophers and political scientists contributing but achieving only intermittent success.

Moved by the political and social upheavals of modernity,

thinkers grew more willing to question whether human perception, ordered by human reason, was the sole metric for making sense of reality. In the late eighteenth and early nineteenth centuries, Romanticism—which was a reaction to the Enlightenment—esteemed human feeling and imagination as true counterparts to reason; it elevated folk traditions, the experience of nature, and a reimagined medieval epoch as preferable to the mechanistic certainties of the modern age.

In the meantime, reason—in the form of advanced theoretical physics—began to progress further toward Kant's thing-in-itself, with disorienting scientific and philosophical consequences. In the late nineteenth and early twentieth centuries, progress at the frontiers of physics began to reveal unexpected aspects of reality. The classical model of physics, whose foundations dated to the early Enlightenment, had posited a world explicable in terms of space, time, matter, and energy, whose properties were in each case absolute and consistent. As scientists sought a clearer explanation for the properties of light, however, they encountered results that this traditional understanding could not explain. The brilliant and iconoclastic theoretical physicist Albert Einstein solved many of these riddles through his pioneering work on quantum physics and his theories of special and general relativity. Yet in doing so, he revealed a picture of physical reality that appeared newly mysterious. Space and time were united as a single phenomenon in which individual perceptions were apparently not bound by the laws of classical physics.[13]

Developing a quantum mechanics to describe this substratum of physical reality, Werner Heisenberg and Niels Bohr challenged long-standing assumptions about the nature of knowledge. Heisenberg emphasized the impossibility of assessing both the position and momentum of a particle accurately and simultaneously. This "uncertainty principle" (as it came to be known) implied that a completely accurate picture of reality might not be available at any given time. Further, Heisenberg argued that physical reality did not have independent inherent form, but was *created* by the process of observation: "I believe that one can formulate the emergence of the classical 'path' of a particle succinctly... *the 'path' comes into being only because we observe it.*"[14]

The question of whether reality had a single true, objective form—and whether human minds could access it—had preoccupied philosophers since Plato. In works such as *Physics and Philosophy: The Revolution in Modern Science* (1958), Heisenberg explored the interplay between the two disciplines and the mysteries that science was now beginning to penetrate. Bohr, in his own pioneering work, stressed that observation affected and ordered reality. In Bohr's telling, the scientific instrument itself—long assumed to be an objective, neutral tool for measuring reality—could never avoid having a physical interaction, however minuscule, with the object of its observation, making it a part of the phenomenon being studied and distorting attempts to describe it. The human mind was forced to choose, among multiple complementary aspects of reality, *which one* it wanted to know accurately at a given moment. A full picture of objective reality, if it were available, could come only by combining impressions of complementary aspects of a phenomenon and accounting for the distortions inherent in each.

These revolutionary ideas penetrated further toward the essence of things than Kant or his followers had thought possible. We are at the beginning of the inquiry into what additional levels of perception or comprehension AI may permit. Its application may allow scientists to fill in gaps in the human observer's ability to measure and perceive phenomena, or in the human (or traditional computer's) ability to process complementary data sets and identify patterns in them.

The twentieth-century philosophical world, jarred by the disjunctions at the frontiers of science and by the First World War, began to chart new paths that diverged from traditional Enlightenment reason and instead embraced the ambiguity and relativity of perception. The Austrian philosopher Ludwig Wittgenstein, who eschewed the academy for life as a gardener and then a village schoolteacher, set aside the notion of a single essence of things identifiable by reason—the goal that philosophers since Plato had sought. Instead, Wittgenstein counseled that knowledge was to be found in generalizations about similarities across phenomena, which he termed "family resemblances": "And the result of this examination is: we see a complicated network of similarities overlapping and criss-crossing: sometimes overall similarities, sometimes similarities of detail." The quest to define and catalog all things, each with its own sharply delineated boundaries, was mistaken, he held. Instead, one should seek to define "This *and similar things*" and achieve familiarity with the resulting concepts, even if they had "blurred" or "indistinct" edges.[15] Later, in the late twentieth century and the early twenty-first, this thinking informed theories of AI and machine learning. Such theories posited that AI's potential lay partly in its ability to scan large data sets to learn types and patterns—e.g., groupings of words often found together, or features most often present in an image when that image was of a cat—and then to make sense of reality by identifying networks of similarities and likenesses with what the AI already knew. Even if AI would never know something in the way a human mind could, an accumulation of matches with the patterns of reality could approximate and sometimes exceed the performance of human perception and reason.

THE ENLIGHTENMENT world—with its optimism regarding human reason despite its consciousness of the pitfalls of flawed human logic—has long been our world. Scientific revolutions, especially in the twentieth century, have evolved technology and philosophy, but the central Enlightenment premise of a knowable world being unearthed, step-by-step, by human minds has persisted. Until now. Throughout three centuries of discovery and exploration, humans have interpreted the world as Kant predicted they would according to the structure of their own minds. But as humans began to approach the limits of their cognitive capacity, they became willing to enlist machines—computers—to augment their thinking in order to transcend those limitations. Computers added a separate digital realm to the physical realm in which humans had always lived. As we are growing increasingly dependent on digital augmentation, we are entering a new epoch in which the reasoning human mind is yielding its pride of place as the sole discoverer, knower, and cataloger of the world's phenomena.

While the technological achievements of the age of reason have been significant, until recently they had remained sporadic enough to be reconciled with tradition. Innovations have been characterized as extensions of previous practices: films were moving photographs, telephones were conversations across space, and automobiles were rapidly moving carriages in which horses were replaced by engines measured by their "horsepower." Likewise, in military life, tanks were sophisticated cavalry, airplanes were advanced artillery, battleships were mobile forts, and aircraft carriers were mobile airstrips. Even nuclear weapons maintained the implication of their moniker—*weapons*—when nuclear powers organized their forces as artillery, emphasizing their prior experience and understanding of war.

But we have reached a tipping point: we can no longer conceive of some of our innovations as extensions of that which we already know. By compressing the time frame in which technology alters the experience of life, the revolution of digitization and the advancement of AI have produced phenomena that are truly new, not simply more powerful or efficient versions of things past. As computers have become faster and smaller, they have become embeddable in phones, watches, utilities, appliances, security systems, vehicles, weapons—and even human bodies. Communication across and between such digital systems is now essentially instantaneous. Tasks that were manual a generation ago—reading, research, shopping, discourse, record keeping, surveillance, and military planning and conduct—are now digital, data-driven, and unfolding in the same realm: cyberspace.[16]

All levels of human organization have been affected by this digitization: through their computers and phones, individuals possess (or at least can access) more information than ever before. Corporations, having become collectors and aggregators of users' data, now wield more power and influence than many sovereign states. Governments, wary of ceding cyberspace to rivals, have entered, explored, and begun to exploit the realm, observing few rules and exercising even fewer restraints. They are quick to designate cyberspace as a domain in which they must innovate in order to prevail over their rivals.

Few have thoroughly understood what exactly has occurred through this digital revolution. Speed is partly to blame, as is inundation. For all its many wondrous achievements, digitization has rendered human thought both less contextual and less conceptual. Digital natives do not feel the need, at least not urgently, to develop concepts that, for most of history, have compensated for the limitations of collective memory. They can (and do) ask search engines whatever they want to know, whether trivial, conceptual, or somewhere in between.

Search engines, in turn, use AI to respond to their queries. In the process, humans delegate aspects of their thinking to technology. But information is not self-explanatory; it is context-dependent. To be useful—or at least meaningful—it must be understood through the lenses of culture and history.

When information is contextualized, it becomes knowledge. When knowledge compels convictions, it becomes wisdom. Yet the internet inundates users with the opinions of thousands, even millions, of other users, depriving them of the solitude required for sustained reflection that, historically, has led to the development of convictions. As solitude diminishes, so, too, does fortitude—not only to develop convictions but also to be faithful to them, particularly when they require the traversing of novel, and thus often lonely, roads. Only convictions—in combination with wisdom—enable people to access and explore new horizons.

The digital world has little patience for wisdom; its values are shaped by approbation, not introspection. It inherently challenges the Enlightenment proposition that reason is the most important element of consciousness. Nullifying restrictions that historically have been imposed on human conduct by distance, time, and language, the digital world proffers that connection, in and of itself, is meaningful.

As online information has exploded, we have turned to software programs to help us sort it, refine it, make assessments based on patterns, and to guide us in answering our questions. The introduction of AI—which completes the sentence we are texting, identifies the book or store we are seeking, and "intuits" articles and entertainment we might enjoy based on prior behavior—has often seemed more mundane than revolutionary. But as it is being applied to more elements of our lives, it is altering the role that our minds have traditionally played in shaping, ordering, and assessing our choices and actions.

CHAPTER 3

FROM TURING TO TODAY — AND BEYOND

IN 1943, WHEN researchers created the first modern computer—electronic, digital, and programmable—their achievement gave new urgency to intriguing questions: Can machines think? Are they intelligent? Could they become intelligent? Such questions seemed particularly perplexing given long-standing dilemmas about the nature of intelligence. In 1950, mathematician and code breaker Alan Turing offered a solution. In a paper unassumingly titled "Computing Machinery and Intelligence," Turing suggested setting aside the problem of machine intelligence entirely. What mattered, Turing posited, was not the mechanism but the *manifestation* of intelligence. Because the inner lives of other beings remain unknowable, he explained, our sole means of measuring intelligence should be external behavior. With this insight, Turing sidestepped centuries of philosophical debate on the nature of intelligence. The "imitation game" he introduced proposed that if a machine operated so proficiently that observers could not distinguish its behavior from a human's, the machine should be labeled intelligent.

The Turing test was born.[1]

Many have interpreted the Turing test literally, imagining robots that pass for people (if that should ever happen) as meeting its criteria. When pragmatically applied, however, the test has proved useful in assessing "intelligent" machines' performance in defined, circumscribed activities such as games. Rather than requiring total indistinguishability from humans, the test applies to machines whose performance is human*like*. In so doing, it focuses on performance, not process. Generators like GPT-3 are AI because they produce text similar to text people produce, not because of the specifics of their models—in GPT-3's case, the fact that it was trained using vast amounts of (online) information.

In 1956, computer scientist John McCarthy further defined artificial intelligence as "machines that can perform tasks that are characteristic of human intelligence." Turing's and McCarthy's assessments of AI have become benchmarks ever since, shifting our focus in defining intelligence to performance (intelligent-seeming *behavior*) rather than the term's deeper philosophical, cognitive, or neuroscientific dimensions.

While for the past half century, machines have largely failed to demonstrate such intelligence, that impasse appears to be at its end. Having operated for decades on the basis of precisely defined code, computers produced analyses that were similarly limited in their rigidity and static nature. Traditional programs could organize volumes of data and execute complex computations but could not identify images of simple objects or adapt to imprecise inputs. The imprecise and conceptual nature of human thought proved to be a stubborn impediment in the development of AI. In the past decade, however, computing innovations have created AIs that have begun to equal or exceed human achievement in such fields.

AIs are imprecise, dynamic, emergent, and capable of "learning." AIs "learn" by consuming data, then drawing observations and conclusions based on the data. While previous systems required exact inputs and outputs, AIs with imprecise function require neither. These AIs translate texts not by swapping individual words but by identifying and employing idiomatic phrases and patterns. Likewise, such AI is considered dynamic because it evolves in response to changing circumstances and emergent because it can identify solutions that are novel to humans. In machinery, these four qualities are revolutionary.

Consider, for example, the breakthrough of AlphaZero in the world of chess. Classical chess programs relied on human expertise, developed by human play, being coded into their programming. But AlphaZero developed its skills by playing millions of games against itself, from which it discovered patterns for itself.

The building blocks of these "learning" techniques are algorithms, sets of steps for translating inputs (such as the rules of a game or measures of quality of moves within those rules) into repeatable outputs (such as winning the game). But machine-learning algorithms are a departure from the precision and predictability of classical algorithms, including those in calculations like long division. Unlike classical algorithms, which consist of steps for producing precise results, machine-learning algorithms consist of steps for improving upon imprecise results. These techniques are making remarkable progress.

Aviation is another example. Soon, AI will pilot or copilot a variety of vehicles in the air. In the DARPA program AlphaDogfight, AI fighter pilots have outperformed humans in simulated combat by executing maneuvers beyond the capabilities of human pilots. Whether piloting jets to fight wars or drones to deliver groceries, AI is poised to have significant impact on the future of both military and civilian aviation.

Although we have seen only the beginnings of such innovations, already, they have subtly altered the fabric of human experience. And in the coming decades, the trend will only accelerate.

Because the technological concepts driving the AI transformation are as complex as they are important, this chapter will be devoted to explaining both the evolution and current state of various types of machine learning and use—both startlingly powerful and inherently limited. A basic introduction to their structure, capabilities, and limitations is vital to understanding the social, cultural, and political changes they have already brought as well as the changes they are likely to produce in the future.

THE EVOLUTION OF AI

Humanity has always dreamed of a helper—a machine capable of performing tasks with the same competence as a human. In Greek mythology, the divine blacksmith Hephaestus forged robots capable of performing human tasks, such as the bronze giant Talos, who patrolled the shores of Crete and protected it from invasion. France's Louis XIV in the seventeenth century and Prussia's Frederick the Great in the eighteenth century harbored a fascination for mechanical automata and oversaw the construction of prototypes. In reality, however, designing a machine and rendering it capable of useful activity—even with the advent of modern computing—has proved devilishly difficult. A central challenge, it turns out, is how—and what—to teach it.

Early attempts to create practically useful AIs explicitly encoded human expertise—via collections of rules or facts—into computer systems. But much of the world is not organized discretely or readily reducible to simple rules or symbolic representations. While in fields that do use precise characterization—chess, algebraic manipulation, and business process automation—AI made great advances, in other fields, like language translation and visual object recognition, inherent ambiguity brought progress to a halt.

The challenges of visual object recognition illustrate the shortcomings of these early programs. Even young children can identify images with ease. But early generations of AI could not. Programmers initially attempted to distill an object's distinguishing characteristics into a symbolic representation. For example, to teach AI to identify a picture of a cat, developers created abstract representations of the various attributes—whiskers, pointy ears, four legs, a body—of an idealized cat. But cats are far from static: they can curl up, run, and stretch, and manifest various sizes and colors. In practice, the approach of formulating abstract models and then attempting to match them with highly variable inputs thereby proved virtually unworkable.

Because these formalistic and inflexible systems were only successful in domains whose tasks could be achieved by encoding clear rules, from the late 1980s through the 1990s, the field entered a period referred to as "AI winter." Applied to more dynamic tasks, AI proved to be brittle, yielding results that failed the Turing test—in other words, that did not achieve or mimic human performance. Because the applications of such systems were limited, R&D funding declined, and progress slowed.

Then, in the 1990s, a breakthrough occurred. At its heart, AI is about performing tasks—about creating machines capable of devising and executing competent solutions to complex problems. Researchers realized that a new approach was required, one that would allow machines to learn on their own. In short, a conceptual shift occurred: we went from attempting to encode human-distilled insights into machines to delegating the learning process itself to the machines.

In the 1990s, a set of renegade researchers set aside many of the earlier era's assumptions, shifting their focus to machine learning. While machine learning dated to the 1950s, new advances enabled practical applications. The methods that have worked best in practice extract patterns from large datasets using neural networks. In philosophical terms, AI's pioneers had turned from the early Enlightenment's focus on reducing the world to mechanistic rules to constructing approximations of reality. To identify an image of a cat, they realized, a machine had to "learn" a range of visual representations of cats by observing the animal in various contexts. To enable machine learning, what mattered was the overlap between various representations of a thing, not its ideal—in philosophical terms, Wittgenstein, not Plato. The modern field of machine learning—of programs that learn through experience—was born.

MODERN AI

Significant progress followed. In the 2000s, in the field of visual object recognition when programmers developed AIs to represent an approximation of an object by learning from a set of images—some of which contained the object, some of which did not—the AIs identified the objects far more effectively than their coded predecessors.

The AI used to identify halicin illustrates the centrality of the machine-learning process. When MIT researchers designed a machine-learning algorithm to predict the antibacterial properties of molecules, training the algorithm with a dataset of more than two thousand molecules, the result was something no conventional algorithm—and no human—could have accomplished. Not only do humans not understand the connections AI revealed between a compound's properties and its antibiotic capabilities, but even more fundamentally, the properties themselves are not amenable to being expressed as rules. A machine-learning algorithm that improves a model based on underlying data, however, is able to recognize relationships that have eluded humans.

As previously noted, such AI is imprecise in that it does not require a predefined relationship between a property and an effect to identify a partial relationship. It can, for example, select highly likely candidates from a larger set of possible candidates. This capability captures one of the vital elements of modern AI. Using machine learning to create and adjust models based on real-world feedback, modern AI can approximate outcomes and analyze ambiguities that would have stymied classical algorithms. Like a classical algorithm, a machine-learning algorithm consists of a sequence of precise steps. But those steps do not directly produce a specific outcome, as they do in a classical algorithm. Rather, modern AI algorithms measure the quality of outcomes and provide means for improving those outcomes, enabling them to be learned rather than directly specified.

Neural networks, inspired by (but, due to complexity, not entirely patterned after) the structure of the human brain, are driving most of these advances. In 1958, Cornell Aeronautical Laboratory researcher Frank Rosenblatt had an idea: Could scientists develop a method for encoding information similar to the method of the human brain, which encodes information by connecting approximately one hundred billion neurons with quadrillions—10^{15}—of synapses? He decided to try: he designed an artificial neural network that encoded relationships between nodes (analogous to neurons) and numerical weights (analogous to synapses). These are networks in that they encode information using a structure of nodes—and the connections between those nodes—in which designated weights represent the strength of the connections between nodes. For decades, a lack of computing power and sophisticated algorithms slowed the development of all but rudimentary neural networks. Advances in both fields, however, have now liberated AI's developers from these restrictions.

In the case of halicin, a neural network captured the association between molecules (the inputs) and their potential to inhibit bacterial growth (the output). The AI that discovered halicin did this without information about chemical processes or drug functions, discovering relationships between the inputs and outputs through deep learning, in which layers of a neural network closer to the input tend to reflect aspects of the input while layers farther from the input tend to reflect broader generalizations that are predictive of the desired output.

Deep learning allows neural networks to capture complex relationships such as those between antibiotic effectiveness and aspects of molecular structure reflected in the training data (atomic weight, chemical composition, types of bonds, and the like). This web allows the AI to capture intricate connections, including connections that can elude humans. In its training phase, as the AI receives new data, it adjusts the weights throughout the network. The network's precision, then, depends on both the volume and quality of the data on which it is trained. As the network receives more data and is composed of more network layers, the weights begin to more accurately capture the relationships. Today's deep networks typically contain around ten layers.

But neural network training is resource-intensive. The process requires substantial computing power and complex algorithms to analyze and adjust to large amounts of data. Unlike humans, most AIs cannot simultaneously train and execute. Rather, they divide their effort into two steps: *training* and *inference.* During the training phase, the AI's quality measurement and improvement algorithms evaluate and amend its model to obtain quality results. In the case of halicin, this was the phase when the AI identified relationships between molecular structures and antibiotic effects based on the training-set data. Then, in the inference phase, researchers tasked the AI with identifying antibiotics that its newly trained model predicted would have a strong antibiotic effect. The AI, then, did not reach conclusions by reasoning as humans reason; it reached conclusions by applying the model it developed.

DIFFERENT TASKS, DIFFERENT LEARNING STYLES

Because the application of AI varies with the tasks it performs, so, too, must the techniques developers use to create that AI. This is a fundamental challenge of deploying machine learning: different goals and functions require different training techniques. But from the combination of various machine-learning methods—notably, the use of neural networks—new possibilities, such as cancer-spotting AIs, emerge.

As of this writing, three forms of machine learning are noteworthy: supervised learning, unsupervised learning, and reinforcement learning. Supervised learning produced the AI that discovered halicin. To recap, when MIT researchers wanted to identify potential new antibiotics, they used a two-thousand-molecule database in order to train a model in which molecular structure was input and antibiotic effectiveness was output. Researchers presented the AI with the molecular structures, each labeled according to its antibiotic effectiveness. Then, given new compounds, the AI estimated the antibiotic effectiveness.

This technique is called supervised learning because the AI developers used a dataset containing example inputs (in this case, molecular structures) that were individually labeled according to the desired output or result (in this case, effectiveness as an antibiotic). Developers have used supervised learning techniques for many purposes, such as creating AIs that recognize images. For this task, the AIs train on a set of prelabeled images, learning to associate an image with its appropriate label—for example, the image of a cat with the label "cat." Having encoded the relationship between images and labels, AIs are then able to correctly identify new images. Therefore, when developers have a dataset that indicates a desired output for each set of inputs, supervised learning has proved to be a particularly effective way of creating a model that can predict outputs in response to novel inputs.

In situations where developers have only troves of data, however, they can employ unsupervised learning to extract potentially useful insights. Thanks to the internet and the digitization of information, businesses, governments, and researchers are awash in data, which they can access more easily than they could in the past. Marketers have more customer information, biologists more DNA data, and bankers more financial transactions on file. When marketers want to identify their customer base, or when fraud analysts seek potential inconsistencies among reams of transactions, unsupervised learning allows AIs to identify patterns or anomalies without having any information regarding outcomes. In unsupervised learning, the training data contains only inputs. Then programmers task the learning algorithm with producing groupings based on some specified weight of measuring the degree of similarity. For example, streaming video services such as Netflix use algorithms to identify clusters of customers with similar viewing habits in order to recommend additional streaming to those customers. But fine-tuning such algorithms can be complex: because most people have several interests, they are typically grouped within several clusters.

AIs trained through unsupervised learning can identify patterns that humans might miss because of the pattern's subtlety, the scale of the data, or both. Because such AIs are trained without specification regarding "proper" outcomes, they can—not unlike the human autodidact—produce surprisingly innovative insights. However, both the human autodidact and these AIs can produce eccentric, nonsensical results.

In both unsupervised and supervised learning, AIs chiefly use data to perform tasks such as discovering trends, identifying images, and making predictions. Looking beyond data analysis, researchers sought to train AIs to operate in dynamic environments. A third major category of machine learning, reinforcement learning, was born.

In reinforcement learning, AI is not passive, identifying relationships within data. Instead, AI is an "agent" in a controlled environment, observing and recording responses to its actions. Generally these are simulated, simplified versions of reality lacking real-world complexities. It is easier to accurately simulate the operation of a robot on an assembly line than it is in the chaos of a crowded city street. But even in a simulated, simplified environment, such as a chess match, a single move can trigger a cascade of opportunities and risks. As a result, directing an AI to train itself in an artificial environment is, in general, insufficient to produce the best performance. Feedback is required.

Providing that feedback is the task of the reward function, indicating to the AI how successful its approach was. No human could effectively fill this role: running on digital processors, AIs can train themselves hundreds, thousands, or billions of times within the space of hours or days, making direct human feedback wholly impractical. Instead, programmers automate such reward functions, carefully specifying precisely how the function operates and the nature of how it simulates reality. Ideally, the simulator provides a realistic experience, and the reward function promotes effective decisions.

AlphaZero's simulator was straightforward: it played against itself. Then, to assess its performance, it employed a reward function[2] that scored its moves based on the opportunities they created. Reinforcement learning requires human involvement in creating the AI training environment (even if not in providing direct feedback during the training itself): humans define a simulator and reward function, and the AI trains itself on that basis. For meaningful results, careful specification of the simulator and the reward function is vital.

THE POWER OF MACHINE LEARNING

From these few building blocks, myriad applications arise. In agriculture, AI is facilitating the precise administration of pesticides, the detection of diseases, and the prediction of crop yields. In medicine, it is facilitating the discovery of new drugs, the identification of new applications of existing drugs, and the detection or prediction of future maladies. (As of this writing, AI has detected breast cancer earlier than human doctors by identifying subtle radiological indicators; it has detected retinopathy, one of the leading causes of blindness, by analyzing retinal photos; it has predicted hypoglycemia in diabetics by analyzing medical histories; and it has detected other heritable conditions by analyzing genetic codes.) In finance, AI is equipped to facilitate high-volume processes: loan approval (or denial), acquisitions, mergers, declarations of bankruptcy, and other transactions.

In other fields, AI is facilitating transcription and translation—in some ways, the most compelling illustration of all. For millennia, humanity has been challenged by the inability of individuals to communicate clearly across cultural and linguistic divides. Mutual miscomprehension, and the inaccessibility of information in one language to a speaker of another, has caused misunderstanding, impeded trade, and fomented war. In the story of the Tower of Babel, it is a symbol of human imperfection—and a bitter penalty for human hubris. Now, it seems, AI is poised to make powerful translation capabilities available to wide audiences, potentially allowing more people to communicate more easily with one another.

Up through the 1990s, researchers attempted to devise rules-based language translation programs. While their efforts had some success in laboratory settings, they failed to yield good results in the real world. The variability and subtlety of language did not reduce to simple rules. All this changed when, in 2015, developers began to apply deep neural networks to the problem. Suddenly, machine translation leaped forward. But its improvement did not just derive from the application of neural networks or machine-learning techniques. Rather, it sprang from new and creative applications of these approaches. These developments underscore a key point: from the basic building blocks of machine learning, developers have the capacity to continue innovating in brilliant ways, unlocking new AIs in the process.

To translate one language to another, a translator needs to capture specific patterns: sequential dependencies. Standard neural networks discern patterns of association between inputs and outputs, such as the sets of chemical properties those antibiotics typically possess. But such networks do not, without modification, capture sequential dependencies, such as the likelihood that a word will appear in a certain position in a sentence given the words that came before it. For example, if a sentence begins with the words "I went to walk the," the next word is far more likely to be *dog* than *cat* or *airplane*. To capture these sequential dependencies, researchers devised networks that use as inputs not only still-to-be-translated text but also text that has already been translated. That way, the AI can identify the next word based on sequential dependencies in the input language *and* in the language the text is being translated to. The most powerful of these networks are *transformers,* which do not need to process language from left to right. Google's BERT, for example, is a bidirectional transformer designed to improve searching.

Additionally, in a considerable shift from conventional supervised learning, language translation researchers employed "parallel corpora," a technique in which specific correspondence between inputs and outputs (for example, meaning between texts in two or more languages) is not needed for training. In conventional approaches, developers trained AI using texts and their preexisting translations—after all, they had the requisite level of correspondence between one language and another. Yet this approach greatly limited the amount of training data as well as the types of text available: although government texts and bestselling books are frequently translated, periodicals, social media, websites, and other informal writings generally are not.

Rather than restricting AIs to training on carefully translated texts, researchers simply supplied articles and other texts in various languages covering a single topic, declining to bother with detailed translations between them. This process, training AIs on roughly matching—but untranslated—bodies of text, is the parallel corpora technique. It is akin to stepping from an introductory language class into a total immersion program. The training is less precise, but the volume of available data is much bigger: developers are able to include news articles, reviews of books and movies, travel stories, and virtually any other formal or informal publication on a topic covered by writers in many languages. The success of this approach has led to more general use of partially supervised learning, in which highly approximate or partial information is used to train.

When Google Translate began to employ deep neural networks trained using parallel corpora, its performance improved by 60 percent—and it has continued to improve ever since.

The radical advancement of automated language translation promises to transform business, diplomacy, media, academia, and other fields as people engage with languages that are not their own more easily, quickly, and cheaply than ever before.

Of course, the ability to translate texts and classify images is one thing. The capacity to generate—to create—new text, images, and sounds is something else. Thus far, the AIs we have described excel at identifying solutions: a chess victory, a drug candidate, a translation good enough to use. But another technique, generative neural networks, can *create.* First, generative neural networks are trained using text or images. Then they produce novel text or images—synthetic but realistic. To illustrate: a standard neural network can identify a picture of a human face, but a *generative* network can create an image of a human face that *seems* real. Conceptually, they depart from their predecessors.

The applications of these so-called generators are staggering. If successfully applied to coding or writing, an author could simply create an outline, leaving the generator to fill in the details. Or an advertiser or filmmaker could supply a generator with a few images or a storyboard, then leave it to the AI to create a synthetic ad or commercial. More concerningly, generators might also be used to create deep fakes—false depictions, indistinguishable from reality, of people doing or saying things they have never done or said. Generators will enrich our information space, but without checks, they will likely also blur the line between reality and fantasy.

A common training technique for the creation of generative AI pits two networks with complementary learning objectives against each other. Such networks are referred to as generative adversarial networks or GANs. The objective of the *generator* network is to create potential outputs, while the objective of the *discriminator* network is to prevent poor outputs from being generated. By analogy, one can think of the generator as being tasked with brainstorming and the discriminator as being tasked with assessing which ideas are relevant and realistic. In the training phase, the generator and discriminator are trained in alternation, holding the generator fixed to train the discriminator and vice versa.

These techniques are not flawless—training GANs can be challenging and often, can produce poor results—but the AIs they yield can achieve remarkable feats. In their most common form, AIs trained with GANs may suggest sentence completions when drafting emails or permit search engines to complete partial queries. More dramatically, GANs may be used to develop AIs that can fill in the details of sketched code—in other words, programmers may soon be able to outline a desired program and then turn that outline over to an AI for completion.

Currently, GPT-3, which can produce human-like text (see chapter 1), is one of the most noteworthy generative AIs. It extends the approach that transformed language translation to language *production.* Given a few words, it can "extrapolate" to produce a sentence, or given a topic sentence, can extrapolate to produce a paragraph. Transformers like GPT-3 detect patterns in sequential elements such as text, enabling them to predict and generate the elements likely to follow. In GPT-3's case, the AI can capture the sequential dependencies between words, paragraphs, or code in order to generate these outputs.

Trained on vast amounts of data drawn primarily from the internet, transformers also can transform text into images and vice versa, expand and condense descriptions, and perform similar tasks. Today, the quality of GPT-3's output—and that of similar AIs—can be impressive but can vary widely. Sometimes, their output appears highly intelligent; at other times, silly or even completely unintelligible. And yet transformers' basic function has the potential to alter many fields, including creative ones. Therefore, they are the subject of considerable interest as researchers and developers probe their strengths, limitations, and applications.

Machine learning has not only broadened the applicability of AI, it has also revolutionized AI even in areas where previous approaches, such as symbolic and rules-based systems, were successful. Machine-learning methods have taken AI from beating human chess experts to discovering entirely new chess strategies. And its capacity for discovery is not limited to games. As we mentioned, DeepMind built an AI that successfully reduced the energy expenditures of Google's data centers by 40 percent more than what its excellent engineers could achieve. This and other advances are taking AI past what Turing envisioned in his test—performance indistinguishable from human intelligence—to include performance that exceeds humans, thereby pushing forward the frontiers of understanding. These advances promise to allow AI to handle new tasks, to make AI more prevalent, and even to allow it to generate original text and code.

Of course, whenever a technology becomes more potent or prevalent, challenges accompany these developments. The personalization of searching—the online function most of us employ most often—is illustrative. In chapter 1, we described the difference between a traditional internet search and an AI-run internet search as the difference between being exposed to designer clothes and being exposed to the full range of clothes available for purchase. An AI enables this outcome—a search engine tailoring itself to an individual user—in two ways: (1) after receiving queries, such as "things to do in New York," an AI can produce *concepts,* such as "walk in Central Park" and "see a show on Broadway," and (2) AI can remember the things a search engine has been asked before and the concepts it has, in response, produced. In addition, it can store these concepts in its version of memory. Over time, it can use its memory to produce concepts that are increasingly specific—and, theoretically, increasingly helpful—to its users. Online streaming services do the same, using AI to make suggestions of television shows and movies "more"—more focused, more positive, or more anything that people want them to be. This can be empowering. AI can steer children away from mature content and, at the same time, toward content appropriate for their ages or frames of reference. AI can steer all of us away from content that is violent, explicit, or otherwise offensive to our sensibilities. It depends on what the algorithms, after analyzing users' past actions, deduce those users' preferences to be. As AI gets to know people, the outcome is largely positive—subscribers to streaming services, for example, become increasingly likely to stream shows and movies that interest rather than offend or confuse them.

The proposition that filtration can help steer choices is both familiar and practical. In the physical world, tourists in foreign countries may hire guides to show them the most historic sites or the most meaningful sites according to their religions, nationalities, or professions. But filtration can become censorship through omission. A guide can avoid slums and high-crime areas. In an authoritarian country, a guide can be a "government minder" and thus only show a tourist what the regime wants him or her to see. But in cyberspace, filtration is self-reinforcing. When the algorithmic logic that personalizes searching and streaming begins to personalize the consumption of news, books, or other sources of information, it amplifies some subjects and sources and, as a practical necessity, omits others completely. The consequence of de facto omission is twofold: it can create personal echo chambers, and it can foment discordance between them. What a person consumes (and thus assumes reflects reality) becomes different from what a second person consumes, and what a second person consumes becomes different still from what a third person consumes—a paradox we consider further in chapter 6.

Managing the risks that increasingly prevalent AI will pose is a task that must be pursued concurrently with the advancement of the field—and it is one of the reasons for this book. We all must pay attention to AI's potential risks. We cannot leave its development or application to any one constituency, be it researchers, companies, governments, or civil society organizations.

AI'S LIMITS AND MANAGEMENT

Unlike earlier generations of AI, in which people distilled a society's understanding of reality in a program's code, contemporary machine-learning AIs largely model reality on their own. While developers may examine the results generated by their AIs, the AIs do not "explain" how or what they learned in human terms. Nor can developers ask an AI to characterize what it has learned. Much as with humans, one cannot really know what has been learned and why (though humans can often offer explanations or justifications that, as of this writing, AI cannot). At best, we can only observe the results an AI produces once it has completed its training. Accordingly, humans must work backward. Once an AI produces a result, people—be they researchers or auditors—must verify that the AI is producing the results desired.

Sometimes, operating beyond the bounds of human experience and unable to conceptualize or generate explanations, AI may produce insights that are true but beyond the frontiers of (at least current) human understanding. When AIs produce unexpected discoveries in this fashion, humans may find themselves in a similar position to that of Alexander Fleming, the discoverer of penicillin. In Fleming's lab, a penicillin-producing mold accidentally colonized a petri dish, killing off disease-causing bacteria and cluing Fleming in to the existence of the potent, previously unknown compound. At the time, humanity, lacking a concept of an antibiotic, did not understand how penicillin worked. The discovery launched an entire field of endeavor. AIs produce similarly startling insights—such as identifying drug candidates and new strategies for winning games—leaving it to humans to divine their significance and, if prudent, integrate these insights into existing bodies of knowledge.

In addition, AI cannot reflect upon what it discovers. Across many eras, humans have experienced war, then reflected on its lessons, its sorrows, and its extremes—from Homer's account of Hector and Achilles at the gates of Troy in *The*

Iliad to Picasso's portrayal of civilian casualties in the Spanish Civil War in *Guernica*. AI cannot do this, nor can it feel the moral or philosophical compulsion to do so. It simply applies its method and produces a result, be that result—from a human perspective—banal or shocking, benign or malignant. AI cannot reflect; the significance of its actions is up to humans to decide. Humans, therefore, must regulate and monitor the technology.

The inability of AI to contextualize or reflect like a human makes its challenges particularly important to attend to. Google's image-recognition software has infamously mislabeled images of people as animals[3] and animals as guns.[4] These errors were plain to any human but eluded the AI. Not only are AIs incapable of reflection, they also make mistakes—including mistakes that any human would regard as rudimentary. And while developers are continually weeding out flaws, deployment has often preceded troubleshooting.

Such misidentifications stem from several sources. Dataset bias is one problem. Machine learning requires data, without which AIs cannot learn good models. A critical problem is that without careful attention, it is more likely that problems of insufficient data will occur for underrepresented groups such as racial minorities. In particular, facial-recognition systems have often been trained on datasets with disproportionately few images of Black people, resulting in poor accuracy. Both the quantity and coverage matter—training AIs on large quantities of highly similar images will result in neural networks that are incorrectly certain of an outcome because they have not encountered it before. In other high-stakes situations, similar underspecification can occur. For example, datasets for training self-driving cars may contain relatively few examples of extraordinary situations, such as when a deer leaps across the road, leaving the AI underspecified as to how to deal with such a scenario. Yet in such scenarios, the AI has to operate at peak levels.

Alternatively, AI bias may result directly from human bias—that is, its training data may contain bias inherent in human actions. This can occur in the labeling of outputs for supervised learning—whatever misidentification the labeler makes, deliberate or inadvertent, the AI will encode. Or a developer may incorrectly specify a reward function used in reinforcement training. Imagine an AI trained to play chess on a simulator that overvalues a set of moves favored by its creator. Like its creator, that AI will learn to prefer those moves, even if they fare poorly in practice.

Of course, the problem of bias in technology is not limited to AI. The pulse oximeter, which has become an increasingly pertinent measurement of two metrics of health—heart rate and oxygen saturation—since the start of the COVID-19 pandemic, overestimates oxygen saturation in dark-skinned individuals. By assuming the way light skin absorbs light is "normal," its designers effectively assumed the way dark skin absorbs light is "abnormal." The pulse oximeter is not run by AI. But still, it fails to pay sufficient attention to a particular population. When AI *is* employed, we should seek to understand its errors—not so we can forgive them but so we can correct them. Bias besets all aspects of human society, and in all aspects of human society, merits a serious response.

Another source of misidentification is rigidity. Consider the case of an animal being misidentified as a gun. The image misleads AIs because it contains subtle characteristics that humans do not detect but AIs can—and can be confused by. AI does not possess what we call common sense. It occasionally conflates two objects that humans could quickly and easily distinguish. Often, what (and how) it conflates is unexpected—not least because, as of this writing, the robustness of AI auditing and compliance regimes is poor. In the real world, an unexpected failure can be more harmful, or at least more challenging, than an expected one: society cannot mitigate what it does not foresee.

AI's brittleness is a reflection of the shallowness of what it learns. Associations between aspects of inputs and outputs based on supervised or reinforcement learning are very different from true human understanding, with its many degrees of conceptualization and experience. The brittleness is also a reflection of AIs' lack of self-awareness. An AI is not sentient. It does not know what it doesn't know. Accordingly, it cannot identify and avoid what to humans might be obvious blunders. This inability of AI to check otherwise clear errors on its own underscores the importance of developing testing that allows humans to identify the limits of an AI's capacities, to review its proposed courses of action, and to predict when an AI is likely to fail.

Accordingly, the development of procedures to assess

whether an AI will perform as expected is vital. Since machine learning will drive AI for the foreseeable future, humans will remain unaware of what an AI is learning and how it knows what it has learned. While this may be disconcerting, it should not be: human learning is often similarly opaque. Artists and athletes, writers and mechanics, parents and children—indeed, all humans—often act on the basis of intuition and thus are unable to articulate what or how they learned. To cope with this opacity, societies have developed myriad professional certification programs, regulations, and laws. Similar techniques should be applied to AIs; for example, societies could permit an AI to be employed only after its creators demonstrate its reliability through testing processes. Developing professional certification, compliance monitoring, and oversight programs for AI—and the auditing expertise their execution will require—will be a crucial societal project.

In industry, pre-use testing exists on a spectrum. App developers often rush programs to market, correcting flaws in real time, while aerospace companies do the opposite: test their jets religiously before a single customer ever sets foot on board. The variance in these regimes depends on several factors—above all, the inherent riskiness of the activity. As AI deployments multiply, the same factors—inherent riskiness, regulatory oversight, market forces—will likely distribute them across the same spectrum, with AIs that drive cars being subjected to significantly greater oversight than AIs that power network platforms for entertainment and connection, such as TikTok.

The division between the learning and inference phases in machine learning permits a testing regime like this to function. When an AI learns continuously, even as it operates, it can develop unexpected or undesirable behavior, as Tay, Microsoft's chatbot, infamously did in 2016. On the internet, Tay encountered hate speech and quickly began to mimic it, forcing its creators to shut it down. Most AIs, though, train in a phase distinct from the operational phase: their learned models—the parameters of their neural networks—are static when they exit training. Because an AI's evolution halts after training, humans can assess its capacities without fear that it will develop unexpected, undesired behaviors after it completes its tests. In other words, when the algorithm is fixed, a self-driving car trained to stop at red lights cannot suddenly "decide" to start running them. This property makes comprehensive testing and certifications possible—engineers may vet a self-driving AI's behavior in a safe environment before uploading it into a car, where an error could cost lives. Of course, fixity does not mean that an AI will not behave unexpectedly when set in novel contexts—it simply means that pretesting is possible. Auditing datasets provides another quality-control check: by ensuring that a facial-recognition AI trains on diverse datasets, or that a chatbot trains on datasets stripped of hate speech, developers can further reduce the risk that the AI will falter when made operational.

As of this writing, AI is constrained by its code in three ways. First, the code sets the parameters of the AI's possible actions. These parameters might be quite broad, permitting a substantial range of autonomy and therefore risk. A self-driving AI can brake, accelerate, and turn, any of which could precipitate a collision. Nevertheless, the parameters of the code establish some limits on the AI's behavior. Though AlphaZero developed novel chess strategies, it did not do so by breaking the rules of chess; it did not suddenly move pawns backward. Actions outside the parameters of the code are beyond the AI's vocabulary. And if the programmer does not put the capacity there, or explicitly forbids the action, the AI cannot do it. Second, AI is constrained by its objective function, which defines and assigns what it is to optimize. In the case of the model that discovered halicin, the objective function was the relationship between the molecules' chemical properties and their antibiotic potential. Limited by its objective function, that AI could not have instead sought to identify molecules that might, for example, help cure cancer. Finally and most obviously, AI can only process inputs that it is designed to recognize and analyze. Without human intervention in the form of an auxiliary program, a translation AI cannot evaluate images—the data would appear nonsensical to it.

One day, AIs may be able to write their own code. For now, efforts to design such AIs are nascent and speculative. Even then, however, AIs would not likely be self-reflective; their objective functions would still define them. They might write code the way AlphaZero plays chess: brilliantly, but without reflection or volition, with strict adherence to the rules.

WHITHER AI

The advances in machine-learning algorithms, combined with increasing data and computational power, have enabled rapid progress in the application of AI, capturing imaginations and investment dollars. The explosion of the research, development, and commercialization of AI, especially machine learning, is global, but it has largely been concentrated in the United States and China.[5] Universities, laboratories, startups, and conglomerates in both countries have been at the forefront of developing and applying machine learning to ever more—and ever more complex—problems.

That said, many aspects of AI and machine learning still need to be developed and understood. Machine-learning-powered AI requires substantial training data. Training data, in turn, requires substantial computing infrastructure, making retraining AI prohibitively expensive, even if it is otherwise desirable to do so. With data and computing requirements limiting the development of more advanced AI, devising training methods that use less data and less computer power is a critical frontier.

Furthermore, despite major advances in machine learning, complex activities that require synthesizing several tasks remain challenging for AI. Driving a car, for example, has proved a formidable challenge, requiring the performance of functions from visual perception to navigation to proactive accident avoidance, all simultaneously. While the field has advanced tremendously over the past decade, driving scenarios vary significantly in terms of how challenging it is to reach human-level performance. Currently, AIs can achieve good performance in structured settings such as limited-access highways and suburban streets with few pedestrians or cyclists. Operating in chaotic settings such as a city's rush-hour traffic, however, remains challenging. Highway driving is particularly interesting, since human drivers in that setting often become bored and distracted, making it possible that AIs could be safer than human drivers for long-distance travel in the not-too-distant future.

Predicting the rate of AI's advance will be difficult. In 1965, engineer Gordon Moore predicted computing power would double every two years—a forecast that has proved remarkably durable. But AI progresses far less predictably. Language-translation AI stagnated for decades, then, through a confluence of techniques and computing power, advanced at a breakneck pace. In just a few years, humans developed AIs with roughly the translation capacity of a bilingual human. How long it will take AI to achieve the qualities of a gifted professional translator—if it ever does—cannot be predicted with precision.

Forecasting how swiftly AI will be applied to additional fields is equally difficult. But we can continue to expect dramatic increases in the capacities of these systems. Whether these advances take five, ten, or twenty-five years, at some point, they will occur. Existing AI applications will become more compact, effective, inexpensive, and, therefore, more frequently used. AI will increasingly become part of our daily lives, both visibly and invisibly.

It is reasonable to expect that over time, AI will progress at least as fast as computing power has, yielding a millionfold increase in fifteen to twenty years. Such progress will allow the creation of neural networks that, in scale, are equal to the human brain. As of this writing, generative transformers have the largest networks. GPT-3 has about 10^{11} such weights. But recently, the state-funded Beijing Academy of Sciences announced a generative language model with 10 times as many weights as GPT-3. This is still 10^4 times fewer than estimates of the human brain's synapses. But if advances proceed at the rate of doubling every two years, this gap could close in less than a decade. Of course, scale does not translate directly to intelligence. Indeed, the level of capability a network will sustain is unknown. Some primates have brains similar in size to or even larger than human brains, but they do not exhibit anything approaching human acumen. Likely, development will yield AI "savants"—programs capable of dramatically exceeding human performance in specific areas, such as advanced scientific fields.

THE DREAM OF ARTIFICIAL GENERAL INTELLIGENCE

Some developers are pushing the frontiers of machine-learning techniques to create what has been dubbed artificial general intelligence (AGI). Like AI, AGI has no precise definition. However, it is generally understood to mean AI capable of completing any intellectual task humans are capable of—in contrast to today's "narrow" AI, which is developed to complete a specific task.

Even more than for current AI, machine learning is critical to the development of AGI, though practical limitations may limit the extent of its expertise to a discrete number of fields, just as the most well-rounded human must still specialize. One possible path to the development of AGI involves training traditional AIs in several fields, then effectively combining their base of expertise into a single AI. Such an AGI might be more well-rounded, able to perform a broader set of activities, and less brittle, blundering less dramatically at the edges of its expertise.

However, scientists and philosophers disagree about whether true AGI is even possible and about what characteristics it might entail. If AGI *is* possible, will it possess the capacities of an average human or of the best human in a given field? In any case, even if AGI could be developed in the manner described above—by combining traditional AIs, training them narrowly and deeply, and gradually agglomerating them to develop a broader base of expertise—it would pose challenges to even the best-funded and most sophisticated researchers. Developing such AIs would require massive computation power and be extremely expensive—with current technology, on the order of billions—so few could afford to create them.

Regardless, it is not obvious that the creation of AGI would substantially alter the trajectory machine-learning algorithms have set humanity on. Whether AI or AGI, human developers will continue to play an important role in creation and operation. The algorithms, training data, and objectives for machine learning are determined by the people developing and training the AI, thus they reflect those people's values, motivations, goals, and judgment. Even as machine-learning techniques become more sophisticated, these limitations will persist.

Whether AI stays narrow or becomes general, it will become more prevalent and more potent. As development and deployment costs decrease, automated devices run by AI will become readily available. Indeed, in conversational interfaces such as Alexa, Siri, and Google Assistant, they already are. Vehicles, tools, and appliances will increasingly be equipped with AIs that automate their activity under our direction and supervision. AIs will be embedded in applications on digital devices and the internet, guiding consumer experiences and revolutionizing enterprises. The world we know will become both more automatic and more interactive (between humans and machines), even if it is not populated with the multipurpose robots of science fiction movies. In the most striking outcomes, human lives will be saved. Self-driving vehicles will reduce auto deaths; other AIs will identify diseases earlier and more precisely; still other AIs will discover drugs and drug-delivery methods in ways that lower research costs—resulting, we hope, in the development of treatments for tenacious maladies and cures for rare diseases. AI aviators will pilot or copilot fleets of delivery drones and even fighter jets. AI coders will complete programs sketched by human developers; AI writers will complete advertisements conceived by human marketers. The efficiency of transportation and logistics will increase, potentially dramatically. AI will reduce energy use and likely find other ways to moderate humans' environmental impact. In the domains of both peace and war, its material effects will be startling.

But its social repercussions are difficult to predict. Consider language translation. Universal translation of spoken language and text will facilitate communication as never before. Such translation will boost commerce and allow unparalleled cross-cultural exchange. And yet this new capability will also carry with it new challenges. Much as social media not only enabled the exchange of ideas but also encouraged polarization, promulgated misinformation, and disseminated hate speech, automated translation may bring languages and cultures together with explosive effects. For centuries, diplomats have carefully managed cross-cultural contact to avoid accidental offense, just as cultural sensitization has often accompanied linguistic training. Instantaneous translation eliminates these buffers. Societies may inadvertently find themselves giving offense and being offended. Will people, relying on automatic translation, exert less effort trying to understand other cultures and nations, increasing their natural tendency to see the world through the lens of their own culture? Or might people become more intrigued by other cultures? Can automatic translation somehow reflect differing cultural histories and sensibilities? Likely there will be no single answer.

The most advanced AIs require vast data, tremendous computing power, and skilled technicians. Unsurprisingly, organizations with access to such resources, both commercial and governmental, drive much of the innovation in this new field. And to leaders, more resources flow. Thus, a cycle of concentration and advancement has defined AI, shaping the experience of individuals, companies, and nations. In many areas, from communication to commerce to security to human consciousness itself, AI will transform our lives and futures. We must all ensure that AI is not created in isolation—and accordingly, pay attention to both its potential benefits and its potential risks.

CHAPTER 4

GLOBAL NETWORK PLATFORMS

Fictional visions of the future of AI technology tend to invoke images of sleek, fully automated self-driving cars and sentient robots that coexist with humans in homes and workplaces, conversing with their users with uncanny intelligence. Inspired by such science fiction scenes, popular conceptions of AI often involve machines that develop a seeming self-awareness, inevitably leading them to misunderstand, decline to obey, or eventually rise up against their human creators. But the anxieties underlying such common fantasies are mistaking the issue by assuming that AI's culmination will be to act like individual humans. We would be better served by recognizing that AI is already all around us—often in ways that are not entirely evident—and redirecting our technological anxieties toward encouraging greater understanding of and transparency regarding AI's integration into our lives.

Social media, web searches, streaming video, navigation, ride sharing, and countless other online services could not operate as they do without the extensive and growing use of AI. By using these online services for the basic activities of daily life—to offer product and service recommendations, select routes, make social connections, and arrive at insights or answers—people around the world are participating in a process that is both mundane and revolutionary. We rely on AI to assist us in pursuing daily tasks without necessarily understanding precisely how or why it is working at any given moment. We are forming new types of relationships that will have substantial implications for individuals, institutions, and nations—between AI and people, between people using AI-facilitated services, and between the creators and operators of these services and governments.

Without significant fanfare—or even visibility—we are integrating nonhuman intelligence into the basic fabric of human activity. This is unfolding rapidly and in connection with a new type of entity we call "network platforms": digital services that provide value to their users by aggregating those users in large numbers, often at a transnational and global scale. In contrast to most products and services, whose value to each user is independent of, or even diminished by, the presence of other users, a network platform's value and attractiveness grows as additional users adopt it—a process economists would label a positive network effect. As more users are drawn to select platforms, such gatherings tend to result in a small number of providers offering a given service, each with a large base of users—sometimes hundreds of millions or even billions. These network platforms increasingly rely on AI, producing an intersection between humans and AI on a scale that suggests an event of civilizational significance.

As AI assumes greater roles on more varied network platforms, these platforms' basic manifestations are becoming material for headlines and geopolitical maneuvers, shaping aspects of individuals' daily reality. Without additional means of explanation, discussion, and oversight that are compatible with a society's values and conducive to some degree of social and political consensus, a rebellion may unfold against the advent of new and seemingly impersonal and inexorable forces—as with the rise of Romanticism in the nineteenth century and the explosion of radical ideologies in the twentieth. Before significant disruption arises, governments, network platform operators, and users must consider the nature of their goals, the basic premises and parameters of their interactions, and the type of world they aim to create.

In less than a generation, the most successful network platforms have brought together user bases larger than the populations of most nations and even some continents. However, large user populations gathered on popular network platforms have more diffuse borders than those of political geography, and network platforms are operated by parties with interests that may differ from those of a nation. Operators of network platforms do not necessarily think in terms of government priorities or national strategy, particularly if those priorities and strategies might conflict with serving their customers. Such network platforms may host or facilitate economic and social interactions that surpass (in number and scale) those of most countries, despite the platforms' having formed no economic or social policy as a government would have. Thus, although they are operated as commercial entities, some network platforms are becoming geopolitically significant actors by virtue of their scale, function, and influence.

Many of the most significant network platforms originated in the United States (Google, Facebook, Uber) or China (Baidu, WeChat, Didi Chuxing). As a result, these network platforms seek to build their user bases and commercial partnerships in regions containing markets that are commercially and strategically significant to Washington and Beijing. Such dynamics introduce novel factors into foreign policy calculations. Commercial competition between network platforms can affect geopolitical competition between governments—sometimes even topping the agenda in diplomacy. This is further complicated by the fact that network platform operators' corporate cultures and strategies are often developed to reflect the priorities of customers and of research and technology hubs, both of which may be far from national capitals.

In countries where they operate, certain network platforms have become integral to individual life, national political discourse, commerce, corporate organization, and even government functions. Their services—even ones that did not exist in any form until recently—now appear indispensable. As an entity without a single direct precedent from prior eras, network platforms sometimes sit in ambiguous relation to rules and expectations that were largely developed in a predigital world.

The question of how network platforms establish community standards—the rules, set by each operator (often administered with the assistance of AI), governing what content is permissible to create and share—provides a crystallizing example of the incongruity between the modern digital space and traditional rules and expectations. Although in principle most network platforms are content-agnostic, in some situations their community standards become as influential as national laws. Content that a network platform and its AI permit or favor may rapidly gain prominence; content they diminish or sometimes even outright prohibit may be relegated to obscurity. Material that is determined to contain disinformation or violate other content standards may effectively be removed from public circulation.

Issues such as these have arisen swiftly in part because network platforms (and their AI) have rapidly expanded in a digital world that transcends geography. These platforms connect large groups of users across space and time—with instantaneously accessible aggregations of data—in a way that few other human creations approximate.[1] To compound the challenge, once AI has been trained, it typically acts faster than the speed of human cognition. These phenomena are inherently neither positive nor negative; they are realities occasioned by the problems human beings seek to solve, the needs we desire to fulfill, and the technology we create to serve our ends. We are experiencing and facilitating changes that require our attention—in thought, culture, politics, and commerce—well beyond the scope of a single human mind or particular product or service.

When the digital world began to expand decades ago, there was no expectation that creators would or should develop a philosophical framework or define their fundamental relationship to national or global interests. After all, such claims generally had not been made on other industries. Instead, society and governments assessed digital products and services in terms of what worked. Engineers sought practical and efficient solutions—connecting users to information and online social spaces, passengers to cars and drivers, and customers to products. There was general excitement about new capabilities and opportunities. There was little demand for predictions about how these virtual solutions might affect the values and behavior of entire societies, such as patterns of vehicle use and traffic congestion with ride sharing or the real-world political and geopolitical alignments of national institutions with social media.

AI-enabled network platforms were created even more recently; with less than a decade of development, we have yet to establish even the basic vocabulary and concepts for an informed debate about this technology—a gap this book seeks to help remedy. Various individuals, corporations, political parties, civic organizations, and governments will inevitably have differing views on the proper operation and regulation of AI-enabled network platforms. What seems intuitive to the software engineer may be perplexing to the political leader or inexplicable to the philosopher. What the consumer welcomes as a convenience the national security official may view as an unacceptable threat or the political leader may reject as out of keeping with national objectives. What one society may embrace as a welcome guarantee another may interpret as a loss of choice or freedom.

The nature and scale of network platforms is bringing the perspectives and priorities of different worlds together in complex alignments, sometimes creating tension and mutual perplexity. In order for individual, national, and international actors to reach informed conclusions about their relationship to AI—and to one another—we must seek a common frame of reference, beginning with establishing terms for informed policy discussions. Even if our understandings differ, we must aim to understand AI-enabled network platforms by assessing their implications for individuals, companies, societies, nations, governments, and regions. We must act urgently on each level.

UNDERSTANDING NETWORK PLATFORMS

Network platforms are inherently large-scale phenomena. One of the defining characteristics of a network platform is that the more people it serves, the more useful and desirable it becomes to users.[2] AI is becoming increasingly important to network platforms that aim to deliver their services at scale, and as a result nearly every internet user today encounters AI, or at least online content shaped by AI, numerous times a day.

For example, Facebook (like many other social networks) has developed increasingly specific community standards for the removal of objectionable content and accounts, listing dozens of categories of prohibited content as of late 2020. Because the platform has billions of active monthly users and billions of daily views,[3] the sheer scale of content monitoring at Facebook is beyond the capabilities of human moderators alone. Despite Facebook reportedly having tens of thousands of people working on content moderation—with the objective of removing offensive content before users see it—the scale is simply such that it cannot be accomplished without AI. Such monitoring needs at Facebook and other companies have driven extensive research and development in an effort to automate text and image analysis by creating increasingly sophisticated machine learning, natural language processing, and computer vision techniques.

For Facebook, the number of removals is currently on the

order of roughly one billion fake accounts and spam posts per quarter as well as tens of millions of pieces of content involving nudity or sexual activity, bullying and harassment, exploitation, hate speech, drugs, and violence. In order to carry out such removals accurately, human-level judgment is often required. Thus, for the most part, Facebook's human operators and users rely on AI to determine which content warrants consumption or review.[4] While only a small fraction of removals are appealed, those that are tend to be automated removals.

Likewise, AI plays a significant role in Google's search engine, but a role that is relatively recent and rapidly evolving. Originally, Google's search engine relied on highly intricate, human-developed algorithms to organize, rank, and guide users toward information. These algorithms amounted to a set of rules for handling potential user queries. Where results didn't prove useful, human developers could adjust them. In 2015, Google's search team moved from using these human-developed algorithms to implementing machine learning. This change led to a watershed moment: incorporating AI has vastly improved the quality and usability of the search engine, making it better able to anticipate questions and organize accurate results. Despite significant improvements in Google's search engine, however, developers had a relatively vague understanding of why searches were producing particular results. Humans can still guide and adjust the search engine, but they may not be able to explain why one particular page is ranked higher than another. To achieve greater convenience and accuracy, human developers have had to willingly forgo a measure of direct understanding.[5]

As these examples illustrate, the leading network platforms increasingly depend on AI to deliver services, fulfill customer expectations, and meet various government requirements. As AI becomes increasingly critical to network platforms' functioning, it is also becoming, gradually and unobtrusively, a sorter and shaper of reality—and, in effect, an actor on the national and global stage.

The potential social, economic, political, and geopolitical influence of each major network platform (and its AI) is substantially augmented by its degree of positive network effects. Positive network effects occur for information-exchange activities in which the value rises with the number of participants. When the value rises in this manner, success tends to produce further success and a greater likelihood of eventual predominance. People naturally gravitate toward existing gatherings, which leads to larger aggregations of users. For a network platform relatively unconstrained by borders, this dynamic leads to a broader, often transnational geographic scope with correspondingly few major competing services.

Positive network effects did not originate with network platforms. Prior to the rise of digital technology, however, the occurrence of such effects was relatively rare. Indeed, for a traditional product or service, an increase in the number of users can easily detract from rather than add to its value. This situation can produce scarcity (for a product or service that is in high demand or sold out), delays (for a product or service that cannot be delivered simultaneously to all the customers who want it), or a loss in the sense of exclusivity that gave a product its initial cachet (e.g., a luxury item that becomes less sought after when it is widely available).

The classic example of positive network effects arises in markets themselves, be they for goods or stocks. Since at least the early seventeenth century, traders of Dutch East India stocks and bonds gathered in Amsterdam, where stock exchanges provided a means for buyers and sellers to arrive at a common valuation in order to trade securities. With the active participation of more buyers and sellers, a stock exchange becomes more useful and valuable to individual participants. Having more participants increases the chances that a transaction will occur and that its valuation will be "accurate," given that the transaction reflects a larger number of individual negotiations between buyers and sellers. Once a stock exchange has gathered a critical mass of users in a given market, it tends to become the first stop for new buyers and sellers—leaving little incentive or opportunity for another exchange to compete by offering precisely the same service.

When traditional telephones were first developed, telephone networks also demonstrated strong positive network effects. For a telephone service reliant on physical wires to connect calls, having a larger number of other subscribers on the same network creates higher value for each subscriber. Thus, in the early days of telephony, there was strong growth for large service providers. In the United States, universality was initially achieved with one very large network operated by AT&T (originally Bell Telephone), interconnected to a number of smaller, largely rural providers. By the 1980s, technological advances permitted telephone service providers to more readily connect with one another, thereby enabling subscribers of new providers to seamlessly reach those on any (domestic) service. These advances facilitated the regulatory breakup of AT&T, demonstrating to customers that value would remain high even without a single large provider. As technology continued to evolve, customers could reach anyone from their phones regardless of their providers, vastly reducing the positive network effect.[6]

There is no inherent reason for the dynamic of positive network effects to stop at national or regional borders—and network platforms often expand across such terrestrial boundaries. Physical distances and national or linguistic differences are rarely obstacles to expansion: the digital world is accessible from anywhere with internet connectivity, and network platforms' services can typically be delivered in several languages. The main limitations on expansion are those put in place by governments or perhaps technological incompatibility (the former sometimes encouraging the latter). Thus, for each type of service, such as social media and video streaming, there are generally a small number of global network platforms, perhaps complemented by local ones. Their users benefit from, and contribute to, a new, as yet poorly understood phenomenon: the operation of nonhuman intelligence at global scale.

COMMUNITY, DAILY LIFE, AND NETWORK PLATFORMS

The digital world has transformed our experience of daily life. As an individual navigates throughout the day, he or she now benefits from, and contributes to, vast shoals of data. The extent of this data and the options for consuming it are too immense and varied for human minds alone to process. The individual comes to rely, often instinctively or subconsciously, on software processes to organize and cull necessary or useful information—selecting news items to view, movies to watch, and music to play based on a combination of previous individual choices and broadly popular selections. Experiencing such automated curation can be so simple and satisfying that it is only noticed in its absence: for example, try reading news in someone else's Facebook feed or browsing movies using someone else's Netflix account.

AI-enabled network platforms have accelerated this integration process and deepened the connections between individuals and our digital technology. With AI designed and trained to intuit and address human questions and goals, a network platform can become a guide to, interpreter for, and record of options that the human mind once managed itself (albeit less efficiently). Network platforms perform these tasks by aggregating information and experiences on a much broader scale than a single human mind or life span can accommodate, allowing them to produce answers and recommendations that can seem uncannily apt. When considering the purchase of winter boots, for example, even the most dedicated individual shopper would never assess hundreds of thousands of national and regional purchases of similar items, consider recent weather trends, factor in the time of year, review his or her comparable prior searches, and investigate shipping patterns before deciding which pair of boots would be the best purchase. An AI, however, might very well assess all these factors.

As a result, individuals often relate to AI-driven network platforms in a manner that they have not, historically, related to other products, services, or machines. As the individual interacts with the AI, and as the AI adapts to the individual's preferences (internet browsing and search queries, travel history, apparent income level, social connections), a kind of tacit partnership begins to form. The individual comes to rely on such platforms to perform a combination of functions that have traditionally been distributed to businesses, governments, and other humans—becoming a combination of postal service, department store, concierge, confessor, and friend.

The relationship between an individual, a network platform, and its other users is a novel combination of intimate bond and remote connection. Already, AI-enabled network platforms review substantial amounts of user data, much of which is personal (such as location, contact information, networks of friends and associates, and financial and health information). Users turn to AI as a guide or facilitator of a personalized experience. The AI's precision and acuity derive from its ability to review and react to an aggregation of hundreds of millions of similar relationships and trillions of similar interactions across space (the geographic breadth of the user base) and time (the aggregation of past uses). The network platform users and its AI enter into a form of compact, interacting with and learning from each other.

At the same time, a network platform's AI follows a logic that is nonhuman and, in many ways, inscrutable to humans. For example, in practice, when an AI-enabled network platform is assessing an image, social media post, or search query, humans may not understand precisely how the AI operates in that particular situation. While Google's engineers know that their AI-enabled search function produced clearer results than it would have without AI, they could not always explain why one particular result was ranked higher than another. To a large extent, AI is judged by the utility of its results, not the process used to reach those results. This signals a shift in priorities from earlier eras, when each step in a mental or mechanical process was either experienced by a human being (a thought, a conversation, an administrative process) or could be paused, inspected, and repeated by human beings.

For example, in much of the industrialized world, the recollection is already fading of an era when travel required "getting directions"—a manual process that might involve an advance phone call to the person being visited, review of a printed city or state map, and, not infrequently, a stop at a gas station or convenience store to validate an assumption or correct a mistake. Now the traveling process unfolds vastly more efficiently through the use of smartphone map applications. Not only can such apps assess several possible routes and the time each would take based on what they "know" about historic traffic patterns at that time of day, they can also factor in accidents and other unique delays on that day (including those that occur during the drive) as well as other indicia (such as other users' searches) that traffic may become worse along a given route during the time the user will take to travel that route.

The shift from atlases to online navigation services has proved so convenient that few have paused to consider what a revolutionary change has occurred or what its consequences might be. The individual and society have gained convenience by entering into a new relationship with a network platform and its operator, accessing and becoming part of an evolving dataset (including tracking of the individual's location, at least while the application is in use) and trusting the network platform and its algorithms to produce accurate results. In a sense, the individual using such a service is not driving alone; instead, he or she is part of a system in which human and machine intelligence are collaborating to guide an aggregation of people through their individual routes.

The prevalence of this type of constant AI companion is likely to increase. As sectors including health care, logistics, retail, finance, communications, media, transportation, and entertainment produce comparable advances—often enabled by network platforms—our experience of day-to-day reality is being transformed.

When users turn to AI-enabled network platforms for assistance with tasks, they are benefiting from a type of gathering and distilling of information that no prior generation has experienced. Such platforms' scale, power, and ability to pursue novel patterns provides individual users with unprecedented conveniences and capabilities. At the same time, these users are entering into a form of human-machine dialogue that has never before existed. AI-enabled network platforms have the capacity to shape human activity in ways that may not be clearly understood—or are even clearly definable or expressible—by the human user. This raises essential questions: With what objective function is such AI operating? And by whose design, and within what regulatory parameters?

The answers to these and similar questions will continue to shape lives and societies in the future: Who operates and defines limits on these processes? What impact might they have on social norms and institutions? And who, if anyone, has access to what AI perceives? If no human can ever fully understand or view the data at an individualized level, or access all the steps involved in the process—that is, if the human role remains confined to designing, monitoring, and setting general parameters for AI—should such limitations be comforting, unnerving, or both?

COMPANIES AND NATIONS

Designers did not set out with the clear objective of inventing AI-enabled network platforms; instead, they arose incidentally, as a function of the problems that individual companies, engineers, and their customers sought to solve. Network platform operators developed their technology to fulfill certain human needs: they connected buyers and sellers, inquirers and information providers, and groups of individuals sharing common interests or goals. They deployed AI to improve—or, increasingly, to enable—their services and to augment their ability to meet users' (and sometimes governments') expectations.

As network platforms have grown and evolved, some have incidentally come to affect activities and sectors of society well beyond their original focus. And, as previously noted, individuals have come to trust certain AI-driven network platforms with information that they would hesitate to show to a friend or the government—such as comprehensive records of where they have gone, what they did (and with whom), and what they searched for and viewed.

The dynamic enabled by access to such personal data puts network platforms, their operators, and the AI they employ in newfound positions of social and political influence. Particularly during a pandemic-influenced era of social distancing and remote work, societies have come to rely on some AI-enabled network platforms as a kind of essential resource and social glue—a facilitator of expression, commerce, food delivery, and transportation. These changes have unfolded at a scale and speed that, thus far, have outpaced a broader understanding and consensus about the roles of these network platforms in society and on the international stage.

As the recent role of social media in conveying and moderating political information and disinformation has demonstrated, some network platforms have assumed functions so significant as to potentially influence the conduct of national governance. This influence has arisen in effect by accident, without necessarily being sought out or properly prepared for. Yet the skills, instincts, and conceptual insights that produce excellence in the world of technology do not inevitably coincide with those of the world of government. Each sphere has its own language, organizational structures, animating principles, and core values. A network platform operating according to its standard commercial objectives and the demands of its users may, in effect, be transcending into the realm of governance and national strategy. In turn, traditional governments may struggle to discern the platform's motives and tactics even as they seek to adjust them to national and global objectives.

The fact that AI operates according to its own processes, which are different from and often faster than human mental processes, adds another complexity. AI develops its own approaches for fulfilling whatever objective functions were specified. It produces outcomes and answers that are not characteristically human and that are largely independent of national or corporate cultures. The global nature of the digital world, and AI's ability to monitor, block, tailor, produce, and distribute information on network platforms worldwide, imports these complexities to the "information space" of disparate societies.

As increasingly sophisticated AI is used to enable network platforms, it shapes social and commercial arrangements on a national and global scale. While social media platforms (and their AI) generally represent themselves as content-agnostic, not only their community standards but also their filtering and presentation of information can influence the way that information is created, aggregated, and perceived. As AI operates to recommend content and connections, categorize information and concepts, and predict user preferences and goals, it may inadvertently reinforce particular individual, group, or societal choices. In effect, it may encourage the distribution of certain types of information and the formation of certain types of connections while discouraging others. This dynamic potentially affects social and political outcomes—regardless of the platform operators' intentions. Each day, individual users and groups influence one another rapidly and at vast scales across countless interactions—particularly when shaped by complex, AI-driven recommendations; as a result, operators may not have a clear understanding of what is occurring in real time. And the complexities are magnified if the operator injects (wittingly or unwittingly) their own values or purposes.

Recognizing these challenges, governmental attempts to address these dynamics will need to proceed with great care.

Any governmental approach to this process—whether to restrict, control, or permit it—necessarily reflects choices and value judgments. If a government encourages platforms to label or block certain content, or if it requires AI to identify and downgrade biased or "false" information, such decisions may effectively operate as engines of social policy with unique breadth and influence. Across the world, the way to address these choices has become the subject of searching debates—particularly in technologically advanced free societies. Any approach is guaranteed to play out on a scale that is vastly greater than nearly any past legal or policy decision—with potentially instantaneous effects on the daily lives of millions or billions of users in many governmental jurisdictions.

The intersection between network platform and governmental arenas will produce unpredictable and, in some cases, highly contested results. Rather than clear outcomes, however, we are more likely to arrive at a series of dilemmas with imperfect answers. Will attempts to regulate network platforms and their AIs function in alignment with various nations' political and social goals (e.g., reducing crime, combating bias) and ultimately produce more just societies? Or will they lead to more powerful and intrusive governments that shape outcomes through a machine proxy, the logic of which is ineffable and the conclusions of which become unavoidable? In iterative exchanges taking place over time, across continents, and between supranational user bases, will AI-driven network platforms advance a shared human culture and quest for answers beyond any national culture or value system? Or will AI-enabled global network platforms amplify specific lessons or patterns divined from users, producing effects that differ from, or even undermine, those their human developers planned or anticipated? We cannot avoid answering these questions because our communications, as now constructed, can no longer operate without AI-assisted networks.

NETWORK PLATFORMS AND DISINFORMATION

National borders have long been permeable to new ideas and trends, including those fostered with a deliberately malign purpose—but never at such scale. While there is broad consensus regarding the importance of preventing intentionally distributed malign disinformation from driving social trends and political events, ensuring this outcome has rarely proved to be a precise or entirely successful undertaking. Moving forward, however, both "offense" and "defense"—both the spread of disinformation and efforts to combat it—will become increasingly automated and entrusted to AI. The language-generating AI GPT-3 has demonstrated the ability to create synthetic personalities, use them to produce language that is characteristic of hate speech, and enter into conversations with human users in order to instill prejudice and even urge them toward violence.[7] If such an AI were to be deployed to spread hate and division at scale, humans alone may not be capable of combating the outcome. Unless such AI is arrested early in its deployment, manually identifying and disabling all its content through individual investigations and decisions would prove deeply challenging for even the most sophisticated governments and network platform operators. For such a vast and arduous task, they would have to turn—as they already do—to content-moderation AI algorithms. But who creates and monitors these and how?

When a free society relies on AI-enabled network platforms that generate, transmit, and filter content across national and regional borders, and when those platforms proceed in a manner that inadvertently promotes hate and division, that society faces a novel threat that should prompt it to consider novel approaches to policing its information environment. The underlying problem is urgent, yet AI-reliant solutions produce their own critical questions. We must not forgo consideration of the proper balance between human judgment and AI-driven automation on both sides of the equation.

For societies accustomed to the free exchange of ideas, grappling with AI's role in assessing and potentially censoring information has introduced difficult fundamental debates. As the tools for spreading disinformation become more powerful and increasingly automated, the process of defining and suppressing disinformation increasingly appears as an essential social and political function. For private corporations and democratic governments, this role brings not only an unusual but also a frequently unsought degree of influence and responsibility over shifts in social and cultural phenomena—developments that previously had not been operated or controlled by any single actor but had developed across millions of individual interactions in the physical world.

For some, the inclination will be to entrust the task to a technical process that *seems* free from human bias and partiality—an AI with an objective function to identify and arrest the flow of disinformation and falsity. But what of the content that is never viewed by the public? When the prominence or diffusion of a message is so curtailed that its existence is, in effect, negated, we have reached a state of censorship. If antidisinformation AI makes a mistake, suppressing content that is not malign disinformation but in fact authentic, how do we identify it? Can we know enough, and in time, to correct it? Alternatively, do we have the right to read, or even a legitimate interest in reading, AI-generated "false" information? The power to train defensive AI against an objective (or subjective) standard of falsehood—and the ability, if any can be developed, to monitor that AI's operations—would in itself become a function of importance and influence rivaling the roles traditionally held by government. Small differences in the design of an AI's objective function, training parameters, and definitions of falsehood could lead to society-altering differences in outcome. These questions become all the more vital as network platforms use AI to provide their services to billions of people.

The international political and regulatory debates over TikTok, an AI-enabled network platform for the creation and sharing of short, often whimsical videos, offers an unexpected early glimpse of the challenges that can arise when relying on AI to shape communications, particularly when that AI is developed in one nation and used by citizens of another. Users of TikTok film and post videos with their smartphones, and many millions of users enjoy watching them. Proprietary AI algorithms recommend content those individuals might enjoy based on their previous use of the platform. Developed in China and having become popular globally, TikTok neither creates content nor appears to set extensive restrictions on it—beyond a time limit on videos and community guidelines that prohibit "misinformation," "violent extremism," and certain types of graphic content.

To the general viewer, the primary attribute of TikTok's AI-assisted lens on the world appears to be whimsicality—its content consists primarily of silly short video snippets of dances, jokes, and unusual skills. Yet because of government concerns about the application's collection of user data and its perceived latent capacity for censorship and disinformation, both the Indian and American governments moved to restrict TikTok's use in 2020. Further, Washington moved to force the sale of TikTok's U.S. operations to a U.S.-based company that could hold user data domestically, preventing it from being exported to China. In turn, Beijing acted to prohibit the exportation of the code that supported the content-recommendation algorithm at the heart of TikTok's efficacy and user appeal.

Soon, more network platforms—perhaps most of those that enable communication, entertainment, commerce, finance, and industrial processes—will rely on increasingly sophisticated, tailor-made AI to deliver key functions and moderate and shape content, often across national borders. The political, legal, and technological ramifications of these maneuvers are still unfolding. That a single AI-enabled whimsical entertainment application has prompted such official multinational consternation suggests that more complex geopolitical and regulatory riddles await us in the near future.

GOVERNMENTS AND REGIONS

Network platforms pose new cultural and geopolitical conundrums not only for individual countries but also, given the natural borderlessness of such technology, for relations between governments and broader regions. Even with substantial and sustained government intervention, most countries—even technologically advanced ones—will not give rise to companies that produce or maintain an advanced "national" version of each globally influential network platform (such as those used for social media, web search, and so on). The pace of technological change is too rapid, and the number of knowledgeable programmers, engineers, and product design and development professionals too few, for such broad coverage. The global demand for talent is too high, the local markets for most services too small, and the product and service costs too substantial to maintain an independent version of each network platform. To stay at the evolving forefront of technological development requires intellectual and financial capital beyond what most companies possess—and beyond what most governments are willing or able to provide. But even in such a scenario, many users, if given the choice, would rather not be limited to a network platform that hosts only their compatriots and the software offerings and content they produce. Instead, the dynamics of positive network effects will tend to support only a handful of participants who are leading the technology and the market for their particular product or service.

Many nations are—and are likely to remain indefinitely—reliant on network platforms that are both designed and hosted in other countries. Thus they are also likely to remain, at least in part, dependent on other countries' regulators for continued access, key inputs, and international updates. Therefore, many governments will have an incentive to guarantee the continued operation of AI-driven online services from other countries that have already been incorporated into fundamental aspects of their society. This undertaking may take the form of regulating network platforms' owners or operators, instituting requirements for their operation, or managing the training of their AI. Governments might insist that developers include steps to avoid certain forms of bias or address particular ethical quandaries.

Public figures may succeed in leveraging a network platform and its AI to obtain greater visibility for their content, enabling them to reach larger audiences. But if platform operators decide that such prominent figures have violated content standards, they can readily be censored or removed, rendering them unable to reach such broad audiences (or driving their audiences underground). Or their content could be accompanied by some form of warning label or other potentially stigmatizing qualification. The issue is what person or institution should make that decision. The authority to independently make and enforce such judgments, now resting with some companies, reflects a level of power that few democratic governments have wielded. While most people would consider it undesirable for private companies to have this degree of power and control, ceding it to government bodies would be almost equally problematic; we have moved beyond conventional policy approaches. When it comes to network platforms, the necessity for such assessments and decisions has arisen swiftly and almost accidentally in recent years, seeming to have surprised users, governments, and companies alike. It needs to be resolved.

NETWORK PLATFORMS AND GEOPOLITICS

The emerging geopolitics of network platforms comprises a key new aspect of international strategy—and governments are not the only players. Governments may increasingly seek to limit the use or behavior of such systems or attempt to prevent them from edging out homegrown rivals in important regions, lest a competing society or economy gain a powerful influence over that country's industrial, economic, or (more difficult to define) political and cultural development. Yet because governments generally do not create or operate these network platforms, the actions of inventors, corporations, and individual users will shape the field along with government restrictions or incentives, creating a strategic arena that is particularly dynamic and difficult to predict. Further, a new form of cultural and political anxiety is being added to this already complex equation. In Beijing, Washington, and some European capitals, concern has been expressed (and articulated obliquely elsewhere) about the implications of conducting broad aspects of national economic and social life on network platforms facilitated by AI designed in other, potentially rival, countries. From this technological and policy ferment, new geopolitical configurations are being established.

The United States has given rise to a globe-spanning, technologically leading set of privately operated network platforms that rely increasingly on AI. The roots of this achievement lie in academic leadership at universities that attract top global talent, a start-up ecosystem that enables participants to bring innovations rapidly to scale and profit from their developments, and government support of advanced R&D (through the National Science Foundation, DARPA, and other agencies). The prevalence of English as a global language, the creation of homegrown or US-influenced technology standards, and the emergence of a substantial domestic base of individual and corporate customers all provide a favorable environment for US network platform operators. Some of these operators eschew government involvement and see their interests as primarily nonnational, while others have embraced government contracts and programs. Abroad, they are all increasingly being treated (often without distinction) as creations and representatives of the United States—although in many cases the US government's role was confined to staying out of their way.

The United States has begun to view network platforms as an aspect of international strategy, restricting the domestic activities of some foreign platforms and restricting the export of some software and technology that could facilitate the growth of foreign competitors. At the same time, federal and state regulators have identified major domestic network platforms as targets for antitrust actions. In the near term, at least, this simultaneous drive for strategic preeminence and domestic multiplicity may push US development in conflicting directions.

China has similarly supported the development of network platforms that are already national in scale, but, at the same time, are poised to expand even further. While Beijing's regulatory approach has encouraged fierce competition among domestic technology players (with global markets as the ultimate goal), it has largely excluded (or mandated heavily tailored offerings by) non-Chinese counterparts within China's borders. In recent years, Beijing has also taken steps to shape international technology standards and bar the export of sensitive domestically developed technologies. Chinese network platforms predominate in China and nearby regions, and some are leading in global markets. Some Chinese network platforms enjoy built-in advantages within Chinese diaspora communities (Chinese-speaking communities in the United States and Europe, for example, continue to heavily use WeChat's financial and messaging functions), but their appeal is not limited to Chinese consumers. Having dominated China's rough-and-tumble domestic market, the country's preeminent network platforms and its AI technology are positioned to compete in the global market.

In certain markets, such as the United States and India, governments have become increasingly outspoken regarding Chinese network platforms (and other Chinese digital technology) as potential or de facto extensions of the Chinese government's policy objectives. While this may be true in certain instances, the difficulties of some Chinese network platform operators suggest that company relationships with the Chinese Communist Party may be complex and varied in practice. Chinese network platform operators may not automatically reflect party or state interests; the correlation is likely to depend on particular network platforms' functions and the extent to which their operators understand and navigate unspoken governmental red lines.

More broadly, while **East and Southeast Asia,** the home of companies with global reach, produce key technologies such as semiconductors, servers, and consumer electronics, they are also the home of locally created network platforms. Across the region, Chinese- and American-hosted platforms are influential to varying degrees among varying segments of the population. In their relationships to network platforms, as in other aspects of economics and geopolitics, the countries of the region have been closely tied to the US-derived technology ecosystem. But there is also substantial use of Chinese network platforms as well as broader engagement with Chinese companies and technology, which East and Southeast Asians may regard as organically connected to their region and integral to their own economic success.

Europe, unlike China and the United States, has yet to create homegrown global network platforms or cultivate the sort of domestic digital technology industry that has supported the development of major platforms elsewhere. Still, Europe commands the attention of the major network platform operators with its leading companies and universities, its tradition of Enlightenment exploration, which laid essential foundations for the computer age, its sizable market, and a regulatory apparatus that is formidable in its ability to innovate and impose legal requirements. Yet Europe continues to face disadvantages for the initial scaling of new network platforms because of its need to serve many languages and national regulatory apparatuses in order to reach its combined market. By contrast, national network platforms in the United States and China are able to start at a continental scale, allowing their companies to better afford the investment needed in order to continue scaling in other languages.

The EU has recently focused regulatory attention on the terms of network platform operators' participation in its market, including these operators' (and other entities') use of AI. As in other geopolitical questions, Europe faces the choice of whether to act as an ally to one side or the other in each major technological sphere—shaping its course by establishing a special relationship—or as a balancer between sides.

Here, the preferences of the traditional EU states and the newer Central and Eastern European entrants may differ, reflecting varying geopolitical and economic situations. Thus far, historic global powers such as France and Germany have prized independence and the freedom to maneuver in their technology policy. However, peripheral European states with recent and direct experience of foreign threats—such as post-Soviet Baltic and Central European states—have shown greater readiness to identify with a US-led "technosphere."

India, while still an emerging force in this arena, has substantial intellectual capital, a relatively innovation-friendly business and academic environment, and a vast reserve of technology and engineering talent that could support the creation of leading network platforms (as has recently been demonstrated with its homegrown online shopping industry). India's population and economy are of a size that could sustain potentially independent network platforms without recourse to other markets. Likewise, Indian-designed network platforms have the potential to become popular in other markets as well. In previous decades, much of India's software talent has been deployed in the IT services industry or in non-Indian network platforms. Now, as the country assesses its regional relationships and relative reliance on imported technology, it may elect either to chart a more independent path or assume a principal role within an international bloc of technologically compatible nations.

Russia, despite a formidable national tradition in math and science, so far has produced few digital products and services with consumer appeal beyond its own borders. Nevertheless, its formidable cyber capabilities and demonstrated ability to penetrate defenses and carry out operations across global networks suggest that Russia must be counted among the important technological powers of the world. Perhaps as a result of exploiting the online vulnerabilities of other countries, Russia has also fostered the use of certain network platforms on a national scale (such as search, e.g., Yandex), though in their present form, these have relatively limited appeal to non-Russian consumers. Currently, these platforms function as a fallback or as an alternative to the dominant providers, not as substantial economic competitors.

Shaped primarily by these governments and regions, a

multidisciplinary contest for economic advantage, digital security, technological primacy, and ethical and social objectives is unfolding—although to date, the principal players have not consistently identified the nature of the contest or the rules of the game.

One approach has been to treat network platforms and their AI as primarily a matter of domestic regulation. In this view, government's principal challenge is to prevent platforms from abusing their positions or shirking previously established or regulated responsibilities. These concepts are evolving and contested, particularly within and between the United States and the EU. And because of the manner in which positive network effects increase value to users with scale, such responsibilities often prove difficult to define.

Another approach has been to treat network platforms' emergence and operations as primarily an issue of international strategy. In this view, the popularization of a foreign operator within a country introduces new cultural, economic, and strategic factors. There is the concern that network platforms may foster, even passively, a level of connection and influence that previously would have arisen only from a close alliance, particularly with the use of AI as a tool for learning from and influencing citizens. If a network platform is useful and successful, it comes to support broader commercial and industrial functions—and, in this capacity, it may become nationally indispensable. At least theoretically, the threatened withdrawal of such a network platform (or its key technological inputs), either by a government or a corporation, serves as a potential instrument of leverage, but by the same token as an incentive to make it dispensable. This hypothetical ability to weaponize network platforms (or other technologies) by withholding them in a crisis may prompt governments to engage in new forms of policy and strategy.

For countries and regions that do not produce homegrown network platforms, the choice for their immediate future seems to be between (1) limiting reliance on platforms that could provide leverage to an adversary government; (2) remaining vulnerable—for example, to another government's potential ability to access data about its citizens; or (3) counterbalancing potential threats against each other. A government may decide that the risks of allowing certain foreign network platforms to operate within its borders are unacceptable—or that they would need to be balanced by the introduction of rival network platforms. Governments with resources may choose to sponsor a domestic entrant as a rival: in many cases, however, this choice would require substantial and sustained intervention—and still risk failure. Advanced countries are likely to try to avoid depending on products of any other single country for key functions (e.g., social media, commerce, ride sharing), particularly in areas where there are several network platforms available globally.

That AI-enabled network platforms created by one society may function and evolve *within* another society and become inextricable from that country's economy and national political discourse marks a fundamental departure from prior eras.

Previously, sources of information and communication were typically local and national in scope—and maintained no independent ability to learn. Today, transportation network platforms created in one country could become the arteries and lifeblood of another country, as the platform learns which consumers need certain products and as it automates the logistics of provision. In effect, such network platforms could become critical economic infrastructure, giving the country of origin leverage over any country that relies on it.

Conversely, when governments elect to limit the reach of foreign technology into their economies, their decisions may hinder that technology's spread—or even its continued commercial viability. Governments may focus on prohibiting the use of foreign network platforms that have been identified as threats. A number of countries have taken such steps for foreign products in general as well as for network platforms in particular. This regulatory approach may create tension with a population's expectation that it should be free to use whatever works best. In open societies, such prohibitions may also raise difficult and novel questions about the proper scope of government regulation.

Caught between governmental actions and concerns regarding their global status and user base, network platform operators will need to make decisions about the extent to which they become, in effect, a conglomeration of national and/or regional companies, potentially in several separate jurisdictions. Conversely, they may decide to conduct themselves as global companies independently pursuing their values, which may not align neatly with any particular government's priorities.

Within the West and China, official assessments of the significance of the other side's digital products and services, including AI-enabled network platforms, have grown. And outside these countries, governments and users may see major network platforms as an expression of American or Chinese culture or interests. Network platform operators' values and organizing principles may mirror those of the society from which they emerged, but in the West, at least, there is no requirement that they correspond. Western corporate cultures often prize self-expression and universality over national interest or conformity to established traditions.

Even where a "technological decoupling" between countries or regions has not occurred, governmental actions are beginning to sort companies into distinctive camps that cater to specific sets of users engaged in particular activities. And as AI learns and adapts to geographically or nationally distinct user bases, it may in turn differently influence human behavior in different regions. In this way, an industry founded on the premise of global community and communication may, in time, facilitate a process of regionalization—uniting blocs of users in separate realities, influenced by distinctive AIs that have evolved in different directions. In time, spheres of regional technology standards could develop, with various AI-enabled network platforms and the activities or expressions they support evolving along parallel but entirely distinct lines and with communication and exchange between them growing increasingly foreign and difficult.

The push and pull of individuals, companies, regulators, and national governments seeking to shape and channel AI-enabled network platforms will grow increasingly complex, conducted alternately as a strategic contest, a trade negotiation, and an ethical debate. Questions that appear urgent may be out of date by the time the relevant official participants have gathered to discuss them. By that time, the AI-enabled network platform may have learned or exhibited new behavior that renders the original terms of the discussion obsolete or insufficient. Creators and operators may come to better understand network platforms' objectives and limits but remain unlikely to intuit probable governmental concerns or broader philosophical objections in advance. Dialogue between these sectors about core concerns and approaches is urgently needed—and should, wherever possible, take place before AI is deployed as part of large-scale network platforms.

AI-ENABLED NETWORK PLATFORMS AND OUR HUMAN FUTURE

Human perception and experience, filtered through reason, has long defined our understanding of reality. This understanding has typically been individual and local in scope, only reaching broader correspondence for certain essential questions and phenomena; it has rarely been global or universal, except in the distinctive context of religion. Now day-to-day reality is accessible on a global scale, across network platforms that unite vast numbers of users. Yet the individual human mind is no longer reality's sole—or perhaps even its principal—navigator. AI-enabled continental and global network platforms have joined the human mind in this task, aiding it and, in some areas, perhaps moving toward eventually displacing it.

New concepts of understanding and limitations—between regions, governments, and network platform operators—must be defined. The human mind has never functioned in the manner in which the internet era demands. With its complex effects on defense, diplomacy, commerce, health care, and transportation posing strategic, technological, and ethical dilemmas too complex for any one actor or discipline to address alone, the advent of AI-enabled network platforms is raising questions that should not be viewed as exclusively national, partisan, or technological in nature.

Strategists need to consider the lessons of prior eras. They should not assume that total victory is possible in each commercial and technological contest. Instead, they should recognize that prevailing requires a definition of success that a society can sustain over time. This, in turn, requires answering the kinds of questions that eluded political leaders and strategic planners during the Cold War era: What margin of superiority will be required? At what point does superiority cease to be meaningful in terms of performance? What degree of inferiority would remain meaningful in a crisis in which each side used its capabilities to the fullest?

Network platform operators will face choices beyond those of serving customers and achieving commercial success. Until now, they have generally not been obliged to define a national or service ethic beyond the organic drive to improve their products, increase their reach, and serve the interests of users and shareholders. As they have assumed broader and more influential roles, however, including functions that influence (and sometimes rival) the activities of governments, they will face far greater challenges. Not only will they need to assist in defining the capacity and ultimate purposes of the virtual realms they have created, they will also need to pay increasing attention to how they interact with one another and with other sectors of society.

CHAPTER 5

SECURITY AND WORLD ORDER

FOR AS LONG as history has been recorded, security has been the minimum objective of an organized society. Cultures have differed in their values, and political units have differed in their interests and aspirations, but no society that could not defend itself—either alone or in alignment with other societies—has endured.

In every era, societies in search of security have sought to turn technological advances into increasingly effective methods of surveilling for threats, achieving superior readiness, exercising influence beyond their borders, and—in the event of war—enabling force in order to prevail. For the earliest organized societies, advances in metallurgy, fortification architecture, horsepower, and shipbuilding were often decisive. In the early modern era, innovations in firearms and cannon, naval vessels, and navigation instruments and techniques played a comparable role. Reflecting on this eternal dynamic in his 1832 classic, *On War,* Prussian military theorist Carl von Clausewitz remarked: "Force, to counter opposing force, equips itself with the inventions of art and science."[1]

Some innovations, such as rampart and moat construction, have favored defense. Yet with each century, a premium has been placed on acquiring means of projecting power across progressively longer distances with progressively greater speed and force. By the time of the American Civil War (1861–65) and the Franco-Prussian War (1870–71), military conflicts had entered the age of the machine, increasingly assuming the characteristics of total war—such as industrialized arms production, orders relayed by telegraph, and troops and materiel transported by rail across continental distances.

With each augmentation of power, major powers have taken one another's measure—assessing which side would prevail in a conflict, what risks and losses such a victory would entail, what would justify them, and how the entry of another power and its arsenal would affect the outcome. The capacities, objectives, and strategies of varied nations were set, at least theoretically, in an equilibrium, or a balance of power.

In the past century, strategy's calibration of means to ends has come out of joint. The technologies used to pursue security have multiplied and grown more destructive, even as the strategies for using them to achieve defined aims have grown more elusive. In our era, the advent of cyber and AI capabilities are adding extraordinary new levels of complexity and abstraction to these calculations.

In this process, World War I (1914–18) was a signal disjunction. In the early 1900s, the major powers of Europe—with advanced economies, pioneering scientific and intellectual communities, and boundless confidence in their global missions—harnessed the technological advances of the Industrial Revolution to construct modern militaries. They accumulated masses of troops by conscription and materiel transportable by train as well as machine guns and other rapid-loading firearms. They developed advanced production methods to replenish arsenals at "machine speed," chemical weapons (whose use has since been outlawed, a ban that most, but not all, governments have accepted), and armored naval vessels and rudimentary tanks. They devised elaborate strategies based on achieving advantage through swift mobilization and alliances based on ironclad pledges among allies to mobilize in concert, swiftly and fully, upon provocation. When a crisis of no inherent global significance arose—the assassination of the heir to the Habsburg throne by a Serbian nationalist—the great powers of Europe followed these plans into a general conflict. The result was a catastrophe that destroyed a generation in pursuit of results that bore no relation to any of the parties' original war aims. Three empires witnessed the collapse of their institutions. Even the victors were depleted for decades and suffered a permanent diminution of their international roles. A combination of diplomatic inflexibility, advanced military technology, and hair-trigger mobilization plans had produced a vicious circle, making global war possible but also unavoidable. Casualties were so enormous that the need to justify them made compromise impossible.

Since that cataclysm, for all the attention, discipline, and resources they have devoted to their arsenals, the major powers have magnified the riddles of modern strategy. At the close of the Second World War and during the opening decades of the Cold War, the two superpowers vied to build nuclear weapons and intercontinental delivery systems—capabilities whose vast destructiveness proved plausibly relatable to only the most grave and total strategic objectives. Observing the first nuclear weapons test in the deserts of New Mexico, the physicist J. Robert Oppenheimer, one of the fathers of the atomic bomb, was moved to invoke not the strategic maxims of Clausewitz but a line from Hindu scripture, the Bhagavad Gita: "Now I am become Death, the destroyer of worlds." This insight presaged the central paradox of Cold War strategy: that the dominant weapons technology of the era was never used. The destructiveness of weapons remained out of proportion to achievable objectives other than pure survival.

The link between capabilities and objectives remained broken throughout the Cold War—or at least not connected in a manner conducive to the clear development of strategy. The major powers constructed technologically advanced militaries and both regional and global alliance systems, but they did not use them against each other or in conflicts with smaller countries or armed movements with more rudimentary arsenals—a bitter truth experienced by France in Algeria, the United States in Korea, and the United States and the Soviet Union in Afghanistan.

THE AGE OF CYBERWARFARE AND AI

Today, after the Cold War, the major powers and other states have augmented their arsenals with cyber capabilities whose utility derives largely from their opacity and deniability and, in some cases, their operation at the ambiguous borders of disinformation, intelligence collection, sabotage, and traditional conflict—creating strategies without acknowledged doctrines. Meanwhile, each advance has been paired with new vulnerabilities.

The AI era risks complicating the riddles of modern strategy further beyond human intention—or perhaps complete human comprehension. Even if nations refrain from the widespread deployment of so-called lethal autonomous weapons—automatic or semiautomatic AI weapons that are trained and authorized to select their own targets and attack without further human authorization—AI holds the prospect of augmenting conventional, nuclear, and cyber capabilities in ways that make security relationships among rivals more challenging to predict and maintain and conflicts more difficult to limit.

AI's potential defensive functions operate on several levels and may soon prove indispensable. Already, AI-piloted fighter jets have shown a substantial ability to dominate human pilots in simulated dogfights. Using some of the same general principles that enabled AlphaZero's victories and the discovery of halicin, AI may identify patterns of conduct that even an adversary did not plan or notice, then recommend methods to counteract them. AI may permit simultaneous translation or the instantaneous relay of other critical information to personnel in crisis zones, whose ability to understand their surroundings or make themselves understood may be essential to a mission or personal safety.

No major country can afford to ignore AI's security dimensions. A race for strategic AI advantage is already taking place, particularly between the United States and China and to some extent Russia.[2] As the knowledge—or suspicion—that others are obtaining certain AI capabilities spreads, more nations will seek them. Once introduced, these capabilities could spread quickly. Although creating a sophisticated AI requires substantial computing power, proliferating or operating the AI generally does not.

The solution to these complexities is neither to despair nor disarm. Nuclear, cyber, and AI technologies exist. Each will inevitably play a role in strategy. None will be "uninvented." If the United States and its allies recoil before the implications of these capabilities and halt progress on them, the result would not be a more peaceful world. Instead, it would be a less balanced world in which the development and use of the most formidable strategic capabilities takes place with less regard for the concepts of democratic accountability and international equilibrium. Both national interest and moral imperative counsel that the United States not cede these fields—indeed, the United States should endeavor to shape them.

Progress and competition in these fields will involve transformations that will test traditional concepts of security. Before these transformations reach a point of inexorability, some effort must be made to define AI-related strategic doctrines and compare them to those of other AI powers (states and nonstate actors alike). In the decades to come, we will need to achieve a balance of power that accounts for the intangibles of cyber conflicts and mass-scale disinformation as well as the distinctive qualities of AI-facilitated war. Realism compels a recognition that AI rivals, even as they compete, should endeavor to explore setting limits on the development and use of exceptionally destructive, destabilizing, and unpredictable AI capabilities. A sober effort at AI arms control is not at odds with national security; it is an attempt to ensure that security is pursued and achieved in the context of a human future.

NUCLEAR WEAPONS AND DETERRENCE

In prior eras, when a new weapon emerged, militaries integrated it into their arsenals and strategists devised doctrines that enabled its use in pursuit of political ends. The advent of nuclear weapons broke this link. The first, and to date only, use of nuclear weapons in war—by the United States against Hiroshima and Nagasaki in 1945, compelling a swift end to the Second World War in the Pacific—was recognized immediately as a watershed. Even as the world's major powers redoubled their efforts to master the new weapons technology and incorporate it into their arsenals, they engaged in unusually open and searching debate about the strategic and moral implications of its use.

With power on a scale far beyond that of any other form of armament at the time, nuclear weapons posed fundamental questions: Could this tremendous destructive force be related, by way of some guiding principle or doctrine, to the traditional elements of strategy? Could the use of nuclear weapons be reconciled with political objectives short of total war and mutual destruction? Would the bomb admit of calibrated, proportional, or tactical use?

The answer, to date, has ranged from ambiguous to negative. Even during the brief period when the United States held a nuclear monopoly (1945 to 1949)—and in the somewhat longer period during which it possessed substantially more effective nuclear delivery systems—it never developed a strategic doctrine or identified a moral principle that persuaded it to use nuclear weapons in an actual conflict following the Second World War. After that, absent clear doctrinal lines that had been mutually agreed upon by the existing nuclear powers—and perhaps not even then—no policy maker could know what would follow a "limited" use and whether it would remain limited. To date, such an attempt has not been made. During a 1955 crisis over shelling across the Taiwan Strait, President Eisenhower—threatening the then nonnuclear People's Republic of China if it did not deescalate—remarked that he saw no reason why tactical nuclear weapons could not be used "just exactly as you would use a bullet or anything else."[3] Nearly seven decades later, no leader has yet tested this proposition.

Instead, during the Cold War, the overriding objective of nuclear strategy became *deterrence*—the use of weapons, primarily through a declared willingness to deploy them, to prevent an adversary from taking action, either by initiating a conflict or using its own nuclear weapons in one. At its core, nuclear deterrence was a psychological strategy of negative objectives. It aimed to persuade an opponent *not* to act by means of a threatened counteraction. This dynamic depended both on a state's physical capacities and on an intangible quality: the potential aggressor's state of mind and its opponent's ability to shape it. Viewed through the lens of deterrence, seeming weakness could have the same consequences as an actual deficiency; a bluff taken seriously could prove a more useful deterrent than a bona fide threat that was ignored. Unique among security strategies (at least until now), nuclear deterrence rests on a series of untestable abstractions: the deterring power could not prove how or by what margin something had been prevented.

Despite these paradoxes, nuclear arsenals were incorporated into basic concepts of international order. When the United

States possessed a nuclear monopoly, its arsenal was used to deter conventional attacks and extend a "nuclear umbrella" over free or allied countries. A Soviet advance across Western Europe was held in check by the prospect, however remote or abstract, that the United States would use nuclear weapons to arrest the attack. Once the Soviet Union crossed the nuclear threshold, the principal purpose of both superpowers' nuclear weapons increasingly became deterring the use of those weapons by the other side. The existence of "survivable" nuclear capabilities—that is, nuclear weapons that could be launched in a counterattack following an adversary's hypothetical first strike—was relied upon to deter nuclear war itself. And it achieved that objective with respect to conflict among the superpowers.

The Cold War hegemons expended tremendous resources on expanding their nuclear capabilities at the same time as their arsenals grew increasingly remote from the day-to-day conduct of strategy. The possession of these arsenals did not deter nonnuclear states—China, Vietnam, Afghanistan—from challenging the superpowers, nor did it stop Central and Eastern Europeans from demanding autonomy from Moscow.

During the Korean War, the Soviet Union was the only nuclear power beyond the United States, and the latter possessed a decisive advantage in the number of weapons and means of delivery. Yet American policy makers refrained from using them, opting to suffer tens of thousands of casualties in World War I–style battles against Soviet-aligned (in retrospect, tenuously) nonnuclear Chinese and North Korean forces rather than embrace the uncertainty or moral opprobrium of nuclear escalation. Since then, every nuclear power confronting a nonnuclear opponent has reached the same conclusion, even when facing defeat at the hands of its nonnuclear foe.

During this era, policy makers did not want for strategies. Under the 1950s doctrine of massive retaliation, the United States threatened to respond to *any* assault, nuclear or conventional, with massive nuclear escalation. Yet a doctrine designed to turn any conflict, however minor, into Armageddon proved psychologically and diplomatically untenable—as well as partially ineffective. In response, some strategists proposed doctrines that would permit the use of tactical nuclear weapons in limited nuclear war.[4] Yet these propositions foundered on concerns regarding escalation and limits. Policy makers feared that the doctrinal lines strategists proposed were too illusory to halt escalation into a global nuclear war. As a result, nuclear strategy remained focused on deterrence and ensuring the credibility of threats, even under apocalyptic conditions beyond those that any human had ever experienced during war. The United States distributed its weapons geographically and constructed a triad (land, sea, and air) of launch capabilities, ensuring that even a surprise first strike by an adversary would not prevent the United States from mounting a devastating response.[5] The Soviets reportedly explored the use of a system designed to be capable, once switched on by human users, of detecting an incoming nuclear strike and disseminating launch orders for a counterattack without further human intervention—an early exploration of the concept of semiautomated warfare involving delegation of certain command functions to a machine.[6]

Strategists in government and academia found the reliance on nuclear strikes without a defensive counterpart disquieting. They explored defensive systems that, at least in theory, would extend policy makers' decisional window during a nuclear standoff, permitting an opportunity to conduct diplomacy—or, at a minimum, to gather more information and correct misinterpretations. Ironically, however, the pursuit of defensive systems only further accelerated the demand for offensive weapons to penetrate defenses on both sides.

As both superpowers' arsenals grew, the possibility of actually deploying nuclear weapons in the service of preventing or punishing the other side's actions came to seem increasingly surreal and incredible—potentially threatening the logic of deterrence itself. The recognition of this nuclear deadlock produced a new doctrine with a name equal parts threat and sardonic recognition: mutual assured destruction or MAD. Because the number of casualties assumed by this theory, which reduced targets while compounding destructiveness, were vast, increasingly, nuclear weapons were confined to the domain of signaling, including increasing the readiness of key systems and units, moving incrementally toward preparations for a nuclear launch, in ways that were meant to be noticed and heeded. But even sending such signals was done sparingly, lest adversaries misinterpret them and unleash global catastrophe. In quest of security, humanity had produced an ultimate weapon and elaborate strategic doctrines to accompany it. The result was a permeating anxiety that such weaponry might ever be used. Arms control was a concept intended to assuage this dilemma.

ARMS CONTROL

Whereas deterrence sought to prevent nuclear war by threatening it, arms control aimed to prevent nuclear war through the limitation or even abolition of the weapons (or categories of weapons) themselves. This approach was paired with nonproliferation: the concept, underpinned by an elaborate set of treaties, technical safeguards, and regulatory and other control mechanisms, that nuclear weapons and the knowledge and technology supporting their construction should be prevented from spreading beyond the nations that already possessed them. Neither arms control nor nonproliferation measures had been attempted on such scale for any previous weapons technology. To date, neither strategy has fully succeeded. Nor has either been pursued in earnest for the major new classes of weapons, cyber and AI, that have been invented in the post–Cold War era. Yet as entrants to the nuclear, cyber, and AI arenas multiply, the arms-control era still holds lessons worthy of consideration.

Following the nuclear brinkmanship and apparently near conflict of the Cuban Missile Crisis (in October of 1962), the then two superpowers, the United States and the Soviet Union, sought to circumscribe nuclear competition through diplomacy. Even as their arsenals grew, and the Chinese, British, and French arsenals entered a calculus of deterrence, Washington and Moscow authorized their negotiators to engage in more substantive arms-control dialogue. Warily, they tested for limits in nuclear weapons counts and capabilities that would be compatible with the maintenance of strategic equilibrium. Eventually, the two sides agreed to limit not only their offensive arsenals but also—following the paradoxical logic of deterrence, in which vulnerability was held to secure peace—their defensive capabilities. The result was the Strategic Arms Limitation agreement and the Anti-Ballistic Missile Treaty, of the 1970s, and eventually the Strategic Arms Reduction Treaty (START), of 1991. In all cases, ceilings placed on offensive weapons preserved the superpowers' capacities to destroy—and thereby presumably to deter—each other while at the same time moderating the arms races inspired by strategies of deterrence.

Although they remained adversaries and continued to spar for strategic advantage, Washington and Moscow both gained a measure of certainty in their calculations via arms-control negotiations. By educating each other about their strategic capabilities, and by agreeing to certain basic limits and verification mechanisms, they both sought to address the fear that the other would suddenly seize an advantage in a nuclear class of weapons in order to strike first.

These initiatives ultimately went beyond aiming for self-restraint to actively discouraging further proliferation. The United States and Russia, in the mid-1960s, originated a multi-commitment, multimechanism regime intended to prohibit all but the original nuclear states from acquiring or possessing nuclear weapons—in exchange for commitments to help other states harness nuclear technology for renewable energy. Such outcomes were facilitated by a distinctive shared sentiment about nuclear weapons—in politics, culture, and in the relationships between individual Cold War leaders—that recognized that a nuclear war between major powers would involve irreversible decisions and unique risks for victor, vanquished, and bystanders alike.

Nuclear weapons presented policy makers with two persistent related riddles: how to define superiority and how to limit inferiority. In an era in which the two superpowers possessed sufficient weaponry to destroy the world many times over, what did superiority mean? Once an arsenal had been built and deployed in a credibly survivable manner, the link between the acquisition of additional weapons, the advantages obtained, and the objectives served became opaque. At the same time, a handful of nations acquired their own modest nuclear arsenals, calculating that they only needed an arsenal sufficient to inflict devastation—not achieve victory—in order to deter attacks.

Nuclear non-use is not an inherently permanent achievement. It is a condition that must be secured by each successive generation of leaders adjusting the deployments and capabilities of their most destructive weapons to a technology evolving at unprecedented speed. This will become particularly challenging as new entrants with varying strategic doctrines and varying attitudes toward the deliberate infliction of civilian casualties seek to develop nuclear capabilities and as equations of deterrence become increasingly diffuse and uncertain. Into this world of unresolved strategic paradoxes, new capabilities and attendant complexities are emerging.

The first is cyber conflict, which has magnified vulnerabilities as well as expanded the field of strategic contests and the variety of options available to participants. The second is AI, which has the capacity to transform conventional, nuclear, and cyber weapons strategy. The emergence of new technology has compounded the dilemmas of nuclear weapons.

CONFLICT IN THE DIGITAL AGE

Throughout history, a nation's political influence has tended to be roughly correlative to its military power and strategic capabilities—its ability, even if exerted primarily through implicit threats, to inflict damage on other societies. Yet an equilibrium based on a calculus of power is not static or self-maintaining; instead, it relies first on a consensus regarding the constituent elements of power and the legitimate bounds of their use. Likewise, maintaining equilibrium requires congruent assessments among all members of the system—especially rivals—regarding states' relative capabilities and intentions as well as of the consequences of aggression. Finally, the preservation of equilibrium requires an actual, and recognized, balance. When a participant in the system enhances its power disproportionately over others, the system will attempt to adjust—either through the organization of countervailing force or the accommodation of a new reality. When the calculation of equilibrium becomes uncertain, or when nations arrive at fundamentally different calculations of relative power, the risk of conflict through miscalculation reaches its height.

In our era, these calculations have entered a new realm of abstraction. This transformation includes so-called cyber weapons, a class of weapons involving dual-use civilian capabilities so that their status as weapons is ambiguous. In some cases, their utility in exercising and augmenting power derives largely from their users' not disclosing their existence or acknowledging their full range of capabilities. Traditionally, parties to a conflict had no difficulty recognizing that a clash had occurred, or recognizing who the belligerents were. Opponents calculated rivals' capabilities and assessed the speed with which their arsenals could be deployed. None of these traditional verities translates directly to the cyber realm.

Conventional and nuclear weapons exist in physical space, where their deployments can be perceived and their capabilities at least roughly calculated. By contrast, cyber weapons derive an important part of their utility from their opacity; their disclosure may effectively degrade some of their capabilities. Their intrusions exploit previously undisclosed flaws in software, obtaining access to a network or system without the authorized user's permission or knowledge. In the contingency of distributed denial-of-service (DDoS) attacks (as on communication systems), a swarm of seemingly valid information requests may be used to overwhelm systems and make them unavailable for their intended use. In such cases, the true sources of the attack may be masked, making it difficult or impossible to determine (at least in the moment) who is attacking. Even one of the most famous instances of cyber-enabled industrial sabotage—the Stuxnet disruption of manufacturing control computers used in Iranian nuclear efforts—has not been formally acknowledged by any government.

Conventional and nuclear weapons are targetable with relative precision, and moral and legal imperatives direct that they target military forces and installations. Cyber weapons can affect computing and communications systems broadly, often hitting civilian systems with particular force. Cyber weapons can also be coopted, modified, and redeployed by other actors for other purposes. In certain respects, this makes cyber weapons akin to biological and chemical weapons, whose effects can spread in unintended and unknown ways. In many cases, they affect large swaths of societies, not just specific targets on a battlefield.[7]

The attributes that lend cyber weapons their utility render the concept of cyber arms control difficult to conceptualize or pursue. Nuclear arms-control negotiators were able to disclose or describe a class of warheads without negating that weapon's function. Cyber arms-control negotiators (which do not yet exist) will need to solve the paradox that discussion of a cyber weapon's capability may be one and the same with its forfeiture (permitting the adversary to patch a vulnerability) or its proliferation (permitting the adversary to copy the code or method of intrusion).

These challenges are made more complex by the ambiguity surrounding key cyber terms and concepts. Various forms of cyber intrusions, online propaganda, and information warfare are called, by various observers in various contexts, "cyber war," "cyberattacks," and in some commentary "an act of war." But this vocabulary is unsettled and sometimes used inconsistently. Some activities, such as intrusions into networks to collect information, may be analogous to traditional intelligence gathering—though at new scales. Other attacks—such as the election-interference campaigns on social media undertaken by Russia and other powers—are a kind of digitized propaganda, disinformation, and political meddling with a larger scope and impact than in previous eras. They are made possible by the expansiveness of the digital technology and network platforms on which these campaigns unfold. Still other cyber actions have the capacity to inflict physical impacts akin to those suffered during traditional hostilities. Uncertainty over the nature, scope, or attribution of a cyber action may render seemingly basic factors a matter of debate—such as whether a conflict has begun, with whom or what the conflict engages, and how far up the escalation ladder the conflict between the parties may be. In that sense, major countries are engaged in a kind of cyber conflict now, though one without a readily definable nature or scope.[8]

A central paradox of our digital age is that the greater a society's digital capacity, the more vulnerable it becomes.

Computers, communications systems, financial markets, universities, hospitals, airlines, and public transit systems—even the mechanics of democratic politics—involve systems that are, to varying degrees, vulnerable to cyber manipulation or attack. As advanced economies integrate digital command-and-control systems into power plants and electricity grids, shift their governmental programs onto large servers and cloud systems, and transfer data into electronic ledgers, their vulnerability to cyberattack multiplies; they present a richer set of targets so that a successful attack could be substantially devastating. Conversely, in the event of a digital disruption, the low-tech state, the terrorist organization, and even individual attackers may assess that they have relatively much less to lose.

The comparatively low cost of cyber capabilities and operations, and the relative deniability that some cyber operations may provide, has encouraged some states to use semiautonomous actors to perform cyber functions. Not unlike the paramilitary groups that pervaded the Balkans on the eve of World War I, these groups may be difficult to control and may engage in provocative activities without official sanction. Compounded by leakers and saboteurs who can neutralize significant portions of a state's cyber capacity or roil its domestic political landscape (even if these activities do not escalate to the level of traditional armed conflict), the speed and unpredictability of the cyber domain and the variety of actors it contains may tempt policy makers into preemptive action in order to forestall a knockout blow.[9]

The speed and ambiguity of the cyber realm have favored offense—and encouraged concepts of "active defense" and "defending forward," which seek to disrupt and preclude attacks.[10] The degree to which cyber deterrence is possible depends in part on what a defender aims to deter and how success is measured. The most effective attacks have usually been those that occur (often without immediate recognition or formal acknowledgment) below the threshold of traditional definitions of armed conflict. No major cyber actor, governmental or nongovernmental, has disclosed the full range of its capabilities or activities—not even to deter actions by others. Strategy and doctrine are evolving uncertainly in a shadow realm, even as new capabilities are emerging. We are at the beginning of a strategic frontier that requires systemic exploration, close collaboration between government and industry to ensure competitive security capabilities, and—in time, and with appropriate safeguards—discussion among major powers concerning limits.

AI AND THE UPHEAVAL IN SECURITY

The destructiveness of nuclear weapons and the mysteries of cyber weapons are increasingly joined by new classes of capabilities that draw on principles of artificial intelligence discussed in previous chapters. Quietly, sometimes tentatively, but with unmistakable momentum, nations are developing and deploying AI that facilitates strategic action across a wide range of military capabilities, with potentially revolutionary effects on security policy.[11]

The introduction of nonhuman logic to military systems and processes will transform strategy. Militaries and security services training or partnering with AI will achieve insights and influence that surprise and occasionally unsettle. These partnerships may negate or decisively reinforce aspects of traditional strategies and tactics. If AI is delegated a measure of control over cyber weapons (offensive or defensive) or physical weapons such as aircraft, it may rapidly conduct functions that humans carry out only with difficulty. AIs such as the US Air Force's ARTUμ have already flown planes and operated radar systems during test flights. In ARTUμ's case, the AI's developers designed it to make "final calls" without human override but limited its capabilities to flying a plane and operating a radar system.[12] Other countries and design teams may exercise less restraint.

In addition to its potentially transformative utility, AI's capacity for autonomy and separate logic generates a layer of incalculability. Most traditional military strategies and tactics have been based on the assumption of a human adversary whose conduct and decision-making calculus fit within a recognizable framework or have been defined by experience and conventional wisdom. Yet an AI piloting an aircraft or scanning for targets follows its own logic, which may be inscrutable to an adversary and unsusceptible to traditional signals and feints—and which will, in most cases, proceed faster than the speed of human thought.

War has always been a realm of uncertainty and contingency, but the entry of AI to this space will introduce new dimensions. Because AIs are dynamic and emergent, even those powers creating or wielding an AI-designed or AI-operated weapon may not know exactly how powerful it is or exactly what it will do in a given situation. How does one develop a strategy—offensive or defensive—for something that perceives aspects of the environment that humans may not, or may not as quickly, and that can learn and change through processes that, in some cases, exceed the pace or range of human thought? If the effects of an AI-assisted weapon depend on the AI's perception during combat—and the conclusions it draws from the phenomena it perceives—can the strategic effects of some weapons be proved only through use? If a competitor trains its AI in silence and secrecy, can leaders know—outside of a conflict—whether they are ahead or behind in an arms race?

In a traditional conflict, the psychology of the adversary is a critical focal point at which strategic action aims. An algorithm knows only its instructions and objectives, not morale or doubt. Because of AI's potential to adapt in response to the phenomena it encounters, when two AI weapons systems are deployed against each other, neither side is likely to have a precise understanding of the results their interaction will generate or their collateral effects. They may discern only imprecisely the other's capabilities and penalties for entering into a conflict. For engineers and builders, these limitations may put premiums on speed, breadth of effects, and endurance—attributes that may make conflicts more intense and widely felt, and above all, more unpredictable.

At the same time, even with AI, a strong defense is the prerequisite of security. Unilateral abandonment of the new technology is precluded by its ubiquity. Yet even as they arm themselves, governments should assess and attempt to explore how the addition of AI logic to the human experience of battle can render war more humane and precise and reflect on the impact on diplomacy and world order.

AI and machine learning will change actors' strategic and tactical options by expanding the capabilities of existing classes of weapons. Not only can AI enable conventional weapons to be targeted more precisely, it can also enable them to be targeted in new and unconventional ways—such as (at least in theory) at a particular individual or object rather than a location.[13] By poring through vast amounts of information, AI cyber weapons can learn how to penetrate defenses without requiring humans to discover software flaws that can be exploited. By the same token, AI can also be used defensively, locating and repairing flaws before they are exploited. But since the attacker can choose the target, AI gives the party on offense an inherent if not insuperable advantage.

If a country faces combat with an adversary that has trained its AI to fly planes, make independent targeting decisions, and fire, what changes in tactics, strategy, or willingness to resort to larger (or even nuclear) weapons will the incorporation of this technology produce?

AI opens new horizons of capabilities in the information space, including in the realm of disinformation. Generative AI can create vast amounts of false but plausible information. AI-facilitated disinformation and psychological warfare, including the use of artificially created personae, pictures, videos, and speech, is poised to produce unsettling new vulnerabilities, particularly for free societies. Widely shared demonstrations have produced seemingly realistic pictures and video of public figures saying things they have never said. In theory, AI could be used to determine the most effective ways of delivering this synthetic AI-generated content to people, tailoring it to their biases and expectations. If a national leader's synthetic image is manipulated by an adversary to foment discord or issue misleading directives, will the public (or even other governments and officials) discern the deception in time?

In contrast to the field of nuclear weapons, no widely shared proscription and no clear concept of deterrence (or of degrees of escalation) attend such uses of AI. AI-assisted weapons both physical and cyber are being prepared by US rivals, and some are reportedly already being used.[14] AI powers are in a position to deploy machines and systems exercising rapid logic and emergent and evolving behavior to attack, defend, surveil, spread disinformation, and identify and disable one another's AI.

As transformative AI capabilities evolve and spread, major nations will, in the absence of verifiable restraints, continue to strive to achieve a superior position.[15] They will assume that proliferation of AI is bound to occur once useful new AI capabilities are introduced. As a result, aided by such technology's dual civilian and military use and its ease of copying and transmission, AI's fundamentals and key innovations will be, in significant measure, public. Where AIs are controlled, controls may prove imperfect, either because advances in technology render them obsolete or because they prove permeable to a determined actor. New users may adapt underlying algorithms for very different aims. A commercial innovation by one society could be adapted for security or information-warfare purposes by another. The most strategically significant aspects of cutting-edge AI development will frequently be adopted by governments to meet their concepts of national interest.

Efforts to conceptualize a cyber balance of power and AI deterrence are in their infancy, if that. Until these concepts are defined, planning will carry an abstract quality. In a conflict, a warring party may seek to overwhelm the will of its enemy through the use, or threatened use, of a weapon whose effects are not well understood.

The most revolutionary and unpredictable effect may occur at the point where AI and human intelligence encounter each other. Historically, countries planning for battle have been able to understand, if imperfectly, their adversaries' doctrines, tactics, and strategic psychology. This has permitted the development of adversarial strategies and tactics as well as a symbolic language of demonstrative military actions, such as intercepting a jet nearing a border or sailing a vessel through a contested waterway. Yet where a military uses AI to plan or target—or even assist dynamically during a patrol or conflict—these familiar concepts and interactions may become newly strange because they will involve communication with, and interpretation of, an intelligence that is unfamiliar in its methods and tactics.

Fundamentally, the shift to AI and AI-assisted weapons and defense systems involves a measure of reliance on—and, in extreme cases, delegation to—an intelligence of considerable analytic potential operating on a fundamentally different experiential paradigm. Such reliance will introduce unknown or poorly understood risks. For this reason, human operators must be involved in and positioned to monitor and control AI actions that have potentially lethal effects. If this human role does not avoid all error, it at least ensures moral agency and accountability.

The deepest challenge, however, may be philosophical. If aspects of strategy come to operate in conceptual and analytical realms that are accessible to AI but not to human reason, they will become opaque—in their processes, reach, and ultimate significance. If policy makers conclude that AI's assistance in scouring the deepest patterns of reality is necessary to understand the capabilities and intentions of adversaries (who may field their own AI) and respond to them in a timely manner, delegation of critical decisions to machines may grow inevitable. Societies are likely to reach differing instinctive limits on what to delegate and what risks and consequences to accept. Major countries should not wait for a crisis to initiate a dialogue about the implications—strategic, doctrinal, and moral—of these evolutions. If they do, their impact is likely to be irreversible. An international attempt to limit these risks is imperative.

MANAGING AI

These issues must be considered and understood before intelligent systems are sent to confront one another. They acquire additional urgency because the strategic use of cyber and AI capabilities implies a broader field for strategic contests. They will extend beyond historic battlefields to, in a sense, anywhere that is connected to a digital network. Digital programs now control a vast and growing realm of physical systems, and an increasing number of these systems—in some cases down to door locks and refrigerators—are networked. This has produced a system of stunning complexity, reach, and vulnerability.

For AI powers, pursuing some form of understanding and mutual restraint is critical. In cases where systems and capabilities are altered easily and relatively undetectably by a change in computer code, each major government may assume that its adversaries are willing to take strategically sensitive

AI research, development, and deployment one step further than what they have publicly acknowledged or even privately pledged. From a purely *technical* perspective, the lines between engaging AI in reconnaissance, targeting, and lethal autonomous action are relatively easily crossed—making a search for mutual restraint and verification systems as difficult as it is imperative.

The quest for reassurance and restraint will have to contend with the dynamic nature of AI. Once they are released into the world, AI-facilitated cyber weapons may be able to adapt and learn well beyond their intended targets; the very capabilities of the weapon might change as AI reacts to its environment. If weapons are able to change in ways different in scope or kind from what their creators anticipated or threatened, calculations of deterrence and escalation may turn illusory. Because of this, the range of activities an AI is capable of undertaking, both at the initial design phase and during the deployment phase, may need to be adjusted so that a human retains the ability to monitor and turn off or redirect a system that has begun to stray. To avoid unexpected and potentially catastrophic outcomes, such restraints must be reciprocal.

Limitations on AI and cyber capabilities will be challenging to define, and proliferation will be difficult to arrest. Capabilities developed and used by major powers have the potential to fall into the hands of terrorists and rogue actors. Likewise, smaller nations that do not possess nuclear weapons and have limited conventional weapons capability have the capacity to wield outsize influence by investing in leading-edge AI and cyber arsenals.

Inevitably, countries will delegate discrete, nonlethal tasks to AI algorithms (some operated by private entities), including the performance of defensive functions that detect and prevent intrusions in cyberspace. The "attack surface" of a digital, highly networked society will be too vast for human operators to defend manually. As many aspects of human life shift online, and as economies continue to digitize, a rogue cyber AI could disrupt whole sectors. Countries, companies, and even individuals should invest in fail-safes to insulate them from such scenarios.

The most extreme form of such protection will involve severing network connections and taking systems off-line. For nations, disconnection could become the ultimate form of defense. Short of such extreme measures, only AI will be capable of performing certain vital cyber defense functions, in part because of the vast extent of cyberspace and the nearly infinite array of possible actions within it. The most significant defensive capabilities in this domain will therefore likely be beyond the reach of all but a few nations.

Beyond AI-enabled defense systems lies the most vexing category of capabilities—lethal autonomous weapons systems—generally understood to include systems that, once activated, can select and engage targets without further human intervention.[16] The key issue in this domain is human oversight and the capability of timely human intervention.

An autonomous system may have a human "on the loop," monitoring its activities passively, or "in the loop," with human authorization required for certain actions. Unless restricted by mutual agreement that is observed and verifiable, the latter form of weapons system may eventually encompass entire strategies and objectives—such as defending a border or achieving a particular outcome against an adversary—and operate without substantial human involvement. In these arenas, it is imperative to ensure an appropriate role for human judgment in overseeing and directing the use of force. Such limitations will have only limited meaning if they are adopted only unilaterally—by one nation or a small group of nations. Governments of technologically advanced countries should explore the challenges of mutual restraint supported by enforceable verification.[17]

AI increases the inherent risk of preemption and premature use escalating into conflict. A country fearing that its adversary is developing automatic capabilities may seek to preempt it: if the attack "succeeds," there may be no way to know whether it was justified. To prevent unintended escalation, major powers should pursue their competition within a framework of verifiable limits. Negotiation should not only focus on moderating an arms race but also making sure that both sides know, in general terms, what the other is doing. But both sides must expect (and plan accordingly) that the other will withhold its most security-sensitive secrets. There will never be complete trust. But as nuclear arms negotiations during the Cold War demonstrated, that does not mean that no measure of understanding can be achieved.

We raise these issues in an effort to define the challenges that AI introduces to strategy. For all their benefits, the treaties (and the accompanying mechanisms of communication, enforcement, and verification) that came to define the nuclear age were not historical inevitabilities. They were the products of human agency and a mutual recognition of peril—and responsibility.

IMPACT ON CIVILIAN AND MILITARY TECHNOLOGIES

Three qualities have traditionally facilitated the separation of military and civilian domains: technological differentiation, concentrated control, and magnitude of effect. Technologies with either exclusively military or exclusively civilian applications are described as differentiated. Concentrated control refers to technologies that a government can easily manage as opposed to technologies that spread easily and thereby escape government control. Finally, the magnitude of effect refers to a technology's destructive potential.

Throughout history, many technologies have been dual-use. Others have spread easily and widely, and some have had tremendous destructive potential. Until now, though, none has been all three: dual-use, easily spread, *and* potentially substantially destructive. The railroads that delivered goods to market were the same that delivered soldiers to battle—but they had no destructive potential. Nuclear technologies are often dual-use and may generate tremendous destructive capacity, but their complicated infrastructure enables relatively secure governmental control. A hunting rifle may be in widespread use and possess both military and civilian applications, but its limited capacity prevents its wielder from inflicting destruction on a strategic level.

AI breaks this paradigm. It is emphatically dual-use. It spreads easily—being, in essence, no more than lines of code: most algorithms (with some noteworthy exceptions) can be run on single computers or small networks, meaning that governments have difficulty controlling the technology by controlling the infrastructure. Finally, AI applications have substantial destructive potential. This relatively unique constellation of qualities, when coupled with the broad range of stakeholders, produces strategic challenges of novel complexity.

AI-enabled weapons may allow adversaries to launch digital assaults with exceptional speed, dramatically accelerating the human capacity to exploit digital vulnerabilities. As such, a state may effectively have no time to evaluate the signs of an incoming attack. Instead, it may need to respond immediately or risk disablement.[18] If a state has the means, it may elect to respond nearly simultaneously, before the attack can occur fully, constructing an AI-enabled system to scan for attacks and empowering it to counterattack.[19] For the opposing side, the reported existence of such a system and the knowledge that it could act without warning may serve as a spur to additional construction and planning, which may include developing parallel technology or one based on different algorithms. Unless care is taken to develop a common concept of limits, the compulsion to act first may overwhelm the need to act wisely—as was the case in the early twentieth century—if indeed humans participate in such decisions at all.

In the stock market, sophisticated so-called quant firms have recognized that AI algorithms can spot market patterns and react with speed that exceeds that of even the ablest trader. Accordingly, such firms have delegated control over certain aspects of their securities trading to these algorithms. In many cases, these algorithmic systems can exceed human profits by a substantial margin. However, they occasionally grossly miscalculate—potentially far beyond the worst human error.

In the financial world, such errors devastate portfolios but do not take lives. In the strategic domain, however, an algorithmic failure analogous to a "flash crash" could be catastrophic. If strategic defense in the digital realm requires tactical offense, if one side errs in its calculations or its actions, an escalatory pattern might be triggered inadvertently.

Attempts to incorporate these new capabilities into a defined concept of strategy and international equilibrium is complicated by the fact that the expertise required for technological preeminence is no longer concentrated exclusively in government. A wide range of actors and institutions participate in shaping technology with strategic implications—from traditional government contractors to individual inventors, entrepreneurs, start-ups, and private research laboratories. Not all will regard their missions as inherently compatible with national objectives as defined by the federal government. A process of mutual education between industry, academia, and government can help bridge this gap and ensure that key principles of AI's strategic implications are understood in a common conceptual framework. Few eras have faced a strategic and technological challenge so complex and with so little consensus about either the nature of the challenge or even the vocabulary necessary for discussing it.

The unresolved challenge of the nuclear age was that humanity developed a technology for which strategists could find no viable operational doctrine. The dilemma of the AI age will be different: its defining technology will be widely acquired, mastered, and employed. The achievement of mutual strategic restraint—or even achieving a common definition of restraint—will be more difficult than ever before, both conceptually and practically.

The management of nuclear weapons, the endeavor of half a century, remains incomplete and fragmentary. Yet the challenge of assessing the nuclear balance was comparatively straightforward. Warheads could be counted, and their yields were known. Conversely, the capabilities of AI are not fixed; they are dynamic. Unlike nuclear weapons, AIs are hard to track: once trained, they may be copied easily and run on relatively small machines. And detecting their presence or verifying their absence is difficult or impossible with the present technology. In this age, deterrence will likely arise from complexity—from the multiplicity of vectors through which an AI-enabled attack is able to travel and from the speed of potential AI responses.

To manage AI, strategists must consider how it might be integrated into a responsible pattern of international relations. Before weapons are deployed, strategists must understand the iterative effect of their use, the potential for escalation, and the avenues for deescalation. A strategy of responsible use, complete with restraining principles, is essential. Policy makers should endeavor to simultaneously address armament, defensive technologies and strategies, together with arms control rather than considering them as chronologically distinct and functionally antagonistic steps. Doctrines must be formulated and decisions must be made in advance of use.

What, then, will be the requirements of restraint? The traditional imposition of restraint on *capability* is an obvious point of departure. During the Cold War, the approach marked some progress, at least symbolically. Some capabilities were restricted (warheads, for example); others (such as categories of intermediate-range missiles) were banned outright. But neither restricting AIs' underlying capabilities nor restricting their number would be wholly compatible with the technology's widespread civilian use and continual evolution. Additional restraints will have to be studied, focusing on AIs' *learning* and *targeting* capabilities.

In a decision that has partly foreseen this challenge, the United States has distinguished between *AI-enabled weapons,* which make human-conducted war more precise, more lethal, and more efficient, and *AI weapons,* which make lethal decisions autonomously from human operators. The United States has declared its aim to restrict use to the first category. It aspires to a world in which no one, not even the United States itself, possesses the second.[20] This distinction is wise. At the same time, the technology's ability to learn and thus evolve could render restrictions on certain capabilities insufficient. Defining the nature and manner of restraint on AI-enabled weapons, and ensuring restraint is mutual, will be critical.

In the nineteenth and twentieth centuries, nations evolved restrictions on certain forms of warfare: the use of chemical weapons, for example, and the disproportionate targeting of civilians. As AI weapons make vast new categories of activities possible, or render old forms of activities newly potent, the nations of the world must make urgent decisions regarding what is compatible with concepts of inherent human dignity and moral agency. Security demands anticipation of what is coming, not merely reaction to what already exists.

The dilemma posed by AI-related weapons technology is that keeping up research and development is essential for national survival; without it we will lose commercial competitiveness and relevance. But the proliferation inherent in the new technology has so far thwarted any attempt at negotiated restraint, even conceptually.

AN OLD QUEST IN A NEW WORLD

Each major technologically advanced country needs to understand that it is on the threshold of a strategic transformation as consequential as the advent of nuclear weapons—but with effects that will be more diverse, diffuse, and unpredictable. Each society that is advancing the frontiers of AI should aim to convene a body at a national level to consider the defense and security aspects of AI and bridge the perspectives of the varied sectors that will shape AI's creation and deployment. This body should be entrusted with two functions: to ensure competitiveness with the rest of the world and, concurrently, to coordinate research on how to prevent or at least limit unwanted escalation or crisis. On this basis, some form of negotiation with allies and adversaries will be essential.

The paradox of an international system is that every power is driven to act—indeed must act—to maximize its own security. Yet to avoid a constant series of crises, each must accept some sense of responsibility for the maintenance of general peace. And this process involves a recognition of limits. The military planner or security official will think (not incorrectly) in terms of worst-case scenarios and prioritize the acquisition of capabilities to meet them. The statesman (who may be one and the same) is obliged to consider how these capabilities will be used and what the world will look like afterward.

In the AI age, long-held strategic logic should be adapted. We will need to overcome, or at least moderate, the drive toward automaticity before catastrophe ensues. We must prevent AIs operating faster than human decision makers from undertaking irretrievable actions with strategic consequences. Defenses will have to be automated without ceding the essential elements of human control. Ambiguity inherent in the domain—combined with the dynamic, emergent qualities of AI and the ease of dissemination—will complicate assessments. In earlier eras, only a handful of great powers or superpowers bore responsibility for restraining their destructive capabilities and avoiding catastrophe. Soon, proliferation may lead to many more actors assuming a similar task.

Leaders of this era can aspire toward six primary tasks in the control of their arsenals, with their broad and dynamic combination of conventional, nuclear, cyber, and AI capabilities.

First, leaders of rival and adversarial nations must be prepared to speak to one another regularly, as their predecessors did during the Cold War, about the forms of war they do not wish to fight. To aid in this effort, Washington and its allies should organize themselves around interests and values that they identify as common, inherent, and inviolable and that encompass the experiences of the generations that came of age at the end of the Cold War or following it.

Second, the unsolved riddles of nuclear strategy must be given new attention and recognized for what they are—one of the great human strategic, technical, and moral challenges. For many decades, memories of a smoldering Hiroshima and Nagasaki forced recognition of nuclear affairs as a unique and grave endeavor. As former secretary of state George Shultz told Congress in 2018, "I fear people have lost that sense of dread." Leaders of countries with nuclear weapons must recognize their responsibility to work together to prevent catastrophe.

Third, leading cyber and AI powers should endeavor to define their doctrines and limits (even if not all aspects of them are publicly announced) and identify points of correspondence between their doctrines and those of rival powers. If deterrence is to predominate over use, peace over conflict, and limited conflict over general conflict, these terms will need to be understood and defined in terms that reflect the distinctive aspects of cyber and AI.

Fourth, nuclear-weapons states should commit to conducting their own internal reviews of their command-and-control and early warning systems. These fail-safe reviews would identify steps to strengthen protections against cyber threats and unauthorized, inadvertent, or accidental use of weapons of mass destruction. These reviews should also include options for precluding cyberattacks on nuclear command-and-control or early warning assets.

Fifth, countries—especially the major technological ones—should create robust and accepted methods of maximizing decision time during periods of heightened tension and in extreme situations. This should be a common conceptual goal, especially among adversaries, that connects both immediate and long-term steps for managing instability and building mutual security. In a crisis, human beings must bear final responsibility for whether advanced weapons are deployed. Especially adversaries should endeavor to agree on a mechanism to ensure that decisions that may prove irrevocable are made at a pace conducive to human thought and deliberation—and survival.[21]

Finally, the major AI powers should consider how to limit continued proliferation of military AI or whether to undertake a systemic nonproliferation effort backed by diplomacy and the threat of force. Who are the aspiring acquirers of the technology that would use it for unacceptable destructive purposes? What specific AI weapons warrant this concern? And who will enforce the redline? The established nuclear powers explored such a concept for nuclear proliferation, with mixed success. If a disruptive and potentially destructive new technology is permitted to transform the militaries of the world's most inveterately hostile or morally unconstrained governments, strategic equilibrium may prove difficult to attain and conflict then uncontrollable.

Due to the dual-use character of most AI technologies, we have a duty to our society to remain at the forefront of research and development. But this will equally oblige us to understand the limits. If a crisis comes, it will be too late to begin discussing these issues. Once employed in a military conflict, the technology's speed all but ensures that it will impose results at a pace faster than diplomacy can unfold. A discussion of cyber and AI weapons among major powers must be undertaken, if only to develop a common vocabulary of strategic concepts and some sense of one another's red-lines. The will to achieve mutual restraint on the most destructive capabilities must not wait for tragedy to arise. As humanity sets out to compete in the creation of new, evolving, and intelligent weapons, history will not forgive a failure to attempt to set limits. In the era of artificial intelligence, the enduring quest for national advantage must be informed by an ethic of human preservation.

CHAPTER 6

AI AND HUMAN IDENTITY

IN AN AGE in which machines increasingly perform tasks only humans used to be capable of, what, then, will constitute our identity as human beings? As previous chapters have explored, AI will expand what we know of reality. It will alter how we communicate, network, and share information. It will transform the doctrines and strategies we develop and deploy. When we no longer explore and shape reality on our own—when we enlist AI as an adjunct to our perceptions and thoughts—how will we come to see ourselves and our role in the world? How will we reconcile AI with concepts like human autonomy and dignity?

In preceding eras, humans have placed themselves at the center of the story. Although most societies recognize human imperfection, they have held that human capacities and experiences constitute a culmination of what mortal beings can aim to achieve in the world. Indeed, they have celebrated individuals who have exemplified pinnacles of the human spirit, illustrating how we wish to see ourselves. These heroes have varied across societies and across eras—leaders, explorers, inventors, martyrs—but they have all embodied aspects of human achievement and, in so doing, human distinctiveness. In the modern age, our veneration of heroes has focused on pioneering exercisers of reason—astronauts, inventors, entrepreneurs, political leaders—who explore and organize our reality.

Now we are entering an era in which AI—a human creation—is increasingly entrusted with tasks that previously would have been performed, or attempted, by human minds. As AI executes these tasks, producing results approximating and sometimes surpassing those of human intelligence, it challenges a defining attribute of what it means to be human. Moreover, AI is capable of learning, evolving, and becoming "better" (according to the objective function it has been given). This dynamic learning permits AI to achieve complex outcomes that were, until now, the preserve of humans and human organizations.

With the rise of AI, the definitions of the human role, human aspiration, and human fulfillment will change. What human qualities will this age celebrate? What will its guiding principles be? To the two traditional ways by which people have known the world, faith and reason, AI adds a third. This shift will test—and, in some instances, transform—our core assumptions about the world and our place in it. Reason not only revolutionized the sciences, it also altered our social lives, our arts, and our faith. Under its scrutiny, the hierarchy of feudalism fell, and democracy, the idea that reasoning people should direct their own governance, rose. Now AI will again test the principles upon which our self-understanding rests.

In an era in which reality can be predicted, approximated, and simulated by an AI that can assess what is relevant to our lives, predict what will come next, and decide what to do, the role of human reason will change. With it, our senses of our individual and societal purposes will change too. In some areas, AI may augment human reason. In others, AI may prompt in humans the feeling of being tangential to the primary process governing a situation. For the driver whose vehicle selects a different lane or route based on an unexplained—indeed, unspoken—calculation, for the person who is extended or denied credit based on an AI-facilitated review, for the job seeker who is asked to interview or not based on a similar process, and for the scholar who is told the most likely answer by an AI model before his or her research has begun in earnest, the experience may prove efficient but not always fulfilling. For humans accustomed to agency, centrality, and a monopoly on complex intelligence, AI will challenge self-perception.

The advances we have considered thus far are illustrations of the many ways in which AI is changing how we interact with the world and thus how we conceive of ourselves and our role in it. AI makes predictions, such as whether a person is likely to have early stage breast cancer; it makes decisions, such as what move to make in chess; it highlights and filters information, such as what movies to watch or what investments to hold; and it generates humanlike text, from sentences to entire paragraphs and documents. As the sophistication of such capabilities increases, they rapidly become what most people consider creative or expert. The fact that AI is able to make certain predictions or decisions, or generate certain material, does not by itself indicate sophistication akin to that of humans. But in many cases, the results are comparable or superior to those previously produced only by humans.

Consider the text that generative models such as GPT-3 are able to create. Nearly any person with a primary education can do a reasonable job of predicting possible completions of a sentence. But writing documents and code, which GPT-3 can do, requires sophisticated skills that humans spend years developing in higher education. Generative models, then, are beginning to challenge our belief that tasks such as sentence completion are distinct from, and simpler than, writing. As generative models improve, AI stands to lead to new perceptions of both the uniqueness and the relative value of human capabilities. Where will that leave us?

With perceptions of reality complementary to humans', AI may emerge as an effective partner for people. In scientific discovery, creative work, software development, and other comparable fields, there can be great benefits to having an interlocutor with a different perception. But this collaboration will require humans to adjust to a world in which our reason is not the only—and perhaps not the most informative—way of knowing or navigating reality. This portends a shift in human experience more significant than any that has occurred for nearly six centuries—since the advent of the movable-type printing press.

Societies have two options: react and adapt piecemeal, or intentionally begin a dialogue, drawing on all elements of human enterprise, aimed at defining AI's role—and, in so doing, defining ours. The former path we will find by default. The latter will require conscious engagement between leaders and philosophers, scientists and humanists, and other groups.

Ultimately, individuals and societies will have to make up their minds which aspects of life to reserve for human intelligence and which to turn over to AI or human-AI collaboration. Human-AI collaboration does not occur between peers. Ultimately, humans both build and direct AI. But as we grow habituated to and reliant on AI, restricting it may become more costly and psychologically challenging or even more technically complicated. Our task will be to understand the transformations that AI brings to human experience, the challenges it presents to human identity, and which aspects of these developments require regulation or counterbalancing by other human commitments. Charting a human future turns on defining a human role in an AI age.

TRANSFORMING HUMAN EXPERIENCE

For some, the experience of AI will be empowering. In most societies, a small but growing cohort understands AI. For these individuals—the people who build it, train it, task it, and regulate it—and for the policy makers and business leaders who have technical advisers at their disposal, the partnership should be gratifying if at times startling. Indeed, in many fields, the experience of surpassing traditional reason through specialized technology, as in the cases of AI's breakthroughs in medicine, biology, chemistry, and physics, will often prove fulfilling.

Those who lack technical knowledge, or participate in AI-managed processes primarily as consumers, will also frequently find these processes gratifying, as in the case of a busy person who can read or check their email while traveling in a self-driving car. Indeed, embedding AI in consumer products will distribute the technology's benefits widely. However, AI will also operate networks and systems that are not designed for any specific individual user's benefit and are beyond any individual user's control. In these cases, encounters with AI may be disconcerting or disempowering, as when AI recommends one individual over others for a desirable promotion or transfer—or encourages or promotes attitudes that challenge or overpower prevailing wisdom.

For managers, the deployment of AI will have many advantages. AI's decisions are often as accurate or more accurate than humans', and with the proper safeguards, may actually be *less* biased. Similarly, AI may be more effective at distributing resources, predicting outcomes, and recommending solutions. Indeed, as generative AI becomes more prevalent, its ability to produce novel text, images, video, and code may even enable it to perform as effectively as its human counterparts in roles typically considered creative (such as drafting documents and creating advertisements). For the entrepreneur offering new products, the administrator wielding new information, and the developer creating increasingly powerful AI, advances in these technologies may enhance senses of agency and choice.

Optimizing the distribution of resources and increasing the accuracy of decision making is good for society, but for the individual, meaning is more often derived from autonomy and the ability to explain outcomes on the basis of some set of actions and principles. Explanations supply meaning and permit purpose; the public recognition and explicit application of moral principles supply justice. But an algorithm does not offer reasons grounded in human experience to explain its conclusions to the general public. Some people, particularly those who understand AI, may find this world intelligible. But others, greater in number, may not understand why AI does what it does, diminishing their sense of autonomy and their ability to ascribe meaning to the world.

As AI transforms the nature of work, it may jeopardize many people's senses of identity, fulfillment, and financial security. Those most affected by such change and potential dislocation will likely hold blue-collar and middle-management jobs that require specific training as well as professional jobs involving review or interpretation of data or drafting of documents in standard forms.[1] While these changes may create not only new efficiencies but also the need for new workers, those who experience dislocation, even if short-term, may derive little consolation from knowing that it is a temporary aspect of a transition that will increase a society's overall quality of life and economic productivity. Some may find themselves freed from drudgery to focus on the more fulfilling elements of their work. Others may find their skills no longer cutting edge or even necessary.

While these challenges are daunting, they are not unprecedented. Previous technological revolutions have displaced or altered work. Inventions such as the mechanical spinning machine displaced laborers and inspired the rise of the Luddites, members of a political movement who sought to ban—or, failing that, to sabotage—new technologies to preserve their old ways of life. The industrialization of agriculture sparked mass migration to the cities. Globalization altered manufacturing and supply chains, and both prompted changes, even unrest, before many societies ultimately absorbed the changes for their overall betterment. Whatever AI's long-term effects prove to be, in the short term, the technology will revolutionize certain economic segments, professions, and identities. Societies need to be ready to supply the displaced not only with alternative sources of income but also with alternative sources of fulfillment.

DECISION MAKING

In the modern age, the standard reaction to a problem has been to seek a solution, sometimes by identifying the human actors responsible for the original deficiency. This view has assigned both responsibility and agency to humans—and both have contributed to our sense of who we are. Now a new actor is entering these equations and may diminish our sense that we are the primary thinkers and movers in a given situation. At times, all of us—whether we create and control AI or just use it—will interact with AI unwittingly or be presented with AI-facilitated answers or outcomes that we did not request. At times, unseen AI may lend the world a magical congeniality, as when stores seemingly anticipate our visits and our whims. At other times, it may produce a Kafkaesque feeling, as when institutions present life-shaping decisions—offers of employment, decisions about car and home loans, or decisions made by security firms or law enforcement—that no single human can explain.

These tensions—between reasoned explanations and opaque decision making, between individuals and large systems, between people with technical knowledge and authority and people without—are not new. What is new is that another intelligence, one that is not human and often inexplicable in terms of human reason, is the source. What is also new is the pervasiveness and scale of this new intelligence. Those who lack knowledge of AI or authority over it may be particularly tempted to reject it. Frustrated by its seeming usurpation of their autonomy or fearful of its additional effects, some may seek to minimize their use of AI and disconnect from social media or other AI-mediated network platforms, shunning its use (at least knowingly) in their daily lives.

Some segments of society may go further, insisting on remaining "physicalists" rather than "virtualists." Like the Amish and the Mennonites, some individuals may reject AI entirely, planting themselves firmly in a world of faith and reason alone. But as AI becomes increasingly prevalent, disconnection will become an increasingly lonely journey. Indeed, even the possibility of disconnection may prove illusory: as society becomes ever more digitized, and AI ever more integrated into governments and products, its reach may prove all but inescapable.

SCIENTIFIC DISCOVERY

The development of scientific understanding often involves a substantial gap between theory and experiment as well as considerable trial and error. With advances in machine learning, we are beginning to see a new paradigm in which models are derived not from a theoretical understanding, as they have been traditionally, but from AI that draws conclusions based on experimental results. This approach necessitates a different expertise from the one that develops theoretical models or conventional computational models. It requires not only a deep understanding of the problem but also the knowledge of which data, and what representation of that data, will be useful for training an AI model to solve it. In the discovery of halicin, for example, the choice of which compounds, and what attributes of those compounds, to input into the model was on the one hand crucial and on the other fortuitous.

The increase in the importance of machine learning to scientific understanding has produced yet another challenge to our views of ourselves and our role(s) in the world. Science has traditionally been a pinnacle amalgam of human-driven expertise, intuition, and insight. In the long-standing interplay between theory and experiment, human ingenuity drives all aspects of scientific inquiry. But AI adds a nonhuman—and divergent-from-human—concept of the world into scientific inquiry, discovery, and understanding. Machine learning is increasingly producing surprising results that prompt new theoretical models and experiments. Just as chess experts have embraced the originally surprising strategies of AlphaZero, interpreting them as a challenge to improve their own understanding of the game, scientists in many disciplines have begun to do the same. Across the biological, chemical, and physical sciences, a hybrid partnership is emerging in which AI is enabling new discoveries that humans are, in response, working to understand and explain.

A striking example of AI enabling broad-based discovery in the biological and chemical sciences is the development of AlphaFold, which used reinforcement learning to create powerful new models of proteins. Proteins are large, complex molecules that play a central role in the structure, function, and regulation of tissues, organs, and processes in biological systems. A protein is made up of hundreds (or thousands) of smaller units called amino acids, which are attached together to form long chains. Because there are twenty different types of amino acids in the formation of proteins, a common way to represent a protein is as a sequence that is hundreds (or thousands) of characters long, in which each character comes from an "alphabet" of twenty characters.

While amino-acid sequences can be quite useful for studying proteins, they fail to capture one critical aspect of those proteins: the three-dimensional structure that is formed by the chain of amino acids. One can think of proteins as complex shapes that need to fit together in three-dimensional space, much like a lock and key, in order for particular biological or chemical outcomes—such as the progression of a disease or its cure—to occur. The structure of a protein can, in some cases, be measured through painstaking experimental methods such as crystallography. But in many cases, the methods distort or destroy the protein, making it impossible to measure the structure. Thus the ability to determine three-dimensional structure from the amino-acid sequence is critical. Since the 1970s, this challenge has been called *protein folding.*

Before 2016, there had not been much progress toward improving the accuracy of protein folding—until a new program, AlphaFold, yielded major progress. As its name implies, AlphaFold was informed by the approach developers took when they taught AlphaZero to play chess. Like AlphaZero, AlphaFold uses reinforcement learning to model proteins without requiring human expertise—in this case, the known protein structures previous approaches relied upon. AlphaFold has more than doubled the accuracy of protein folding from around 40 to around 85 percent, enabling biologists and chemists around the world to revisit old questions they had been unable to answer and to ask new questions about battling pathogens in people, animals, and plants.[2] Advances like AlphaFold—impossible without AI—are transcending previous limits in measurement and prediction. The result is changes in how scientists approach what they can learn in order to cure diseases, protect the environment, and solve other essential challenges.

EDUCATION AND LIFELONG LEARNING

Coming of age in the presence of AI will alter our relationships, both with one another and with ourselves. Just as a divide exists today between "digital natives" and prior generations, so, too, will a divide emerge between "AI natives" and the people who precede them. In the future, children may grow up with AI assistants, more advanced than Alexas and Google Homes, that will be many things at once: babysitter, tutor, adviser, friend. Such an assistant will be able to teach children virtually any language or train children in any subject, calibrating its style to individual students' performance and learning styles to bring out their best. AI may serve as a playmate when a child is bored and as a monitor when a child's parent is away. As AI-provided and tailored education is introduced, the average human's capabilities stand both to increase and to be challenged.

The boundary between humans and AI is strikingly porous. If children acquire digital assistants at an early age, they will become habituated to them. At the same time, digital assistants will evolve with their owners, internalizing their preferences and biases as they mature. A digital assistant tasked to maximize a human partner's convenience or fulfillment through personalization may produce recommendations and information that are deemed essential even if the human user cannot explain exactly why they are better than any alternative resources.

Over time, individuals may come to prefer their digital assistants over humans, for humans will be less intuitive of their preferences and more "disagreeable" (if only because humans have personalities and desires not keyed to other individuals). As a result, our dependence on one another, on human relationships, may decrease. What, then, will become of the ineffable qualities and lessons of childhood? How will the omnipresent companionship of a machine, which does not feel or experience human emotion (but may mimic it), affect a child's perception of the world and his or her socialization? How will it shape imagination? How will it change the nature of play? How will it alter the process of making friends or fitting in?

Arguably, the availability of digital information has already transformed the education and cultural experience of a generation. Now the world is embarking on another great experiment, in which children will grow up with machines that will, in many ways, act as human teachers have for generations—but without human sensibilities, insight, and emotion. Eventually, the experiment's participants will likely ask whether their experiences are being altered in ways they did not expect or accept.

Parents, alarmed by the potentially uncertain effects of such exposure on their children, may push back. Just as parents a generation ago limited television time and parents today limit screen time, parents in the future may limit AI time. But those who want to push their children to succeed, or who lack the inclination or ability to replace AI with a human parent or tutor—or who simply want to satisfy their children's desire to have AI friends—may sanction AI companionship for their children. So children—learning, evolving, impressionable—may form their impressions of the world in dialogue with AIs.

The irony is that even as digitization is making an increasing amount of information available, it is diminishing the space required for deep, concentrated thought. Today's near-constant stream of media increases the cost, and thus decreases the frequency, of contemplation. Algorithms promote what seizes attention in response to the human desire for stimulation—and what seizes attention is often the dramatic, the surprising, and the emotional. Whether an individual can find space in this environment for careful thought is one matter. Another is that the now-dominant forms of communication are non-conducive to the promotion of tempered reasoning.

NEW INFORMATION INTERMEDIARIES

As we said in chapter 4, AI increasingly shapes our informational domain. To inform and organize human experience, intermediaries have been created—organizations and institutions that distill complex information, highlight what individuals need to know, and broadcast the results.[3] As societies increasingly divided their physical labor, they also divided their mental labor, creating newspapers and journals to inform citizens generally and founding universities to educate them specifically. Since then, information has been aggregated, distilled, and broadcast—and its meaning defined—by such institutions.

Now, in every domain characterized by intensive intellectual labor, from finance to law, AI is being integrated into the process of learning. But humans cannot always verify that what AI presents is representative; we cannot always explain why applications such as TikTok and YouTube promote some videos over others. Human editors and anchors, on the other hand, can provide explanation (accurate or not) of their reasons for selecting what they present. As long as people desire such explanation, the age of AI will disappoint the majority of people who do not understand the technology's processes and mechanisms.

AI's effects on human knowledge are paradoxical. On the one hand, AI intermediaries can navigate and analyze bodies of data vaster than the unaided human mind could have previously contemplated. On the other, this power—the ability to engage with vast bodies of data—may also accentuate forms of manipulation and error. AI is capable of exploiting human passions more effectively than traditional propaganda. Having tailored itself to individual preferences and instincts, AI elicits responses its creator or user desires. Similarly, the deployment of AI intermediaries may also amplify inherent biases, even if these AI intermediaries are technically under human control. The dynamics of market competition prompt social media platforms and search engines to present information that users find most compelling. As a result, information that users are believed to want to see is prioritized, distorting a representative picture of reality. Much as technology accelerated the speed of information production and dissemination in the nineteenth and twentieth centuries, in this era, information is being altered by the mapping of AI onto dissemination processes.

Some people will seek information filters that do not distort, or at least distort transparently. Some will balance filter against filter, independently weighing the results. Others may opt out entirely, preferring filtration by traditional human intermediaries. Yet when the majority of people in a society accept AI intermediation, either as a default or as the price of powering network platforms, those pursuing traditional forms of personal inquiry through research and reason may find themselves unable to keep pace with events. They will certainly find their ability to shape them progressively limited.

If information and entertainment become immersive, personalized, and synthetic—such as AI-sorted "news" confirming people's long-held beliefs or AI-generated movies "starring" long-deceased actors—will a society have a common understanding of its history and current affairs? Will it have a common culture? If an AI is instructed to scan a century's worth of music or television and produce "a hit," does it create or merely assemble? How will writers, actors, artists, and other creators, whose labors have traditionally been treated as a unique human engagement with reality and lived experience, see themselves and be seen by others?

A NEW HUMAN FUTURE

Traditional reason and faith will persist in the age of AI, but their nature and scope are bound to be profoundly affected by the introduction of a new, powerful, machine-operated form of logic. Human identity may continue to rest on the pinnacle of animate intelligence, but human reason will cease to describe the full sweep of the intelligence that works to comprehend reality. To make sense of our place in this world, our emphasis may need to shift from the centrality of human reason to the centrality of human dignity and autonomy.

The Enlightenment was characterized by attempts to define human reason and understand it in relation to, and in contrast with, previous human eras. The political philosophers of the Enlightenment—Hobbes, Locke, Rousseau, and many others—derived their concepts from theoretical states of nature, from which they articulated views of the attributes of human beings and the structure of society. In turn, leaders asked how human knowledge could be pooled and objectively disseminated to permit enlightened government and human flourishing. Absent similarly comprehensive efforts to understand human nature, the disorientations of the AI age are going to prove difficult to mitigate.

The cautious may seek to restrict AI, confining its use to discrete functions and circumscribing when, where, and how it is used. Societies or individuals may reserve the role of principal and judge for themselves, relegating AI to the position of support staff. However, competitive dynamics will challenge limitations, of which the security dilemmas presented in the previous chapter are the starkest evidence. Barring fundamental ethical or legal constraints, what company would forgo knowledge of AI functionality a rival has used to offer new products or services? If AI enables a bureaucrat, architect, or investor to predict outcomes or conclusions with ease, on what basis would he or she not use it? Given the pressures for deployment, limitations on AI uses that are, on their face, desirable will need to be formulated at a society-wide or international level.

AI may take a leading role in exploring and managing

both the physical and digital worlds. In specific domains, humans may defer to AI, preferring its processes to the limitations of the human mind. This deference could prompt many or even most humans to retreat into individual, filtered, customized worlds. In this scenario, AI's power—combined with its prevalence, invisibility, and opacity—will raise questions about the prospects for free societies and even for free will.

In many arenas, AI and humans will instead become equal partners in the enterprise of exploration. Consequently, human identity will come to reflect reconciliation with new relationships, both with AI and with reality. Societies will carve out distinct spheres for human leadership. At the same time, they will develop the social structures and habits necessary to understand and interact fruitfully with AI. Societies need to build the intellectual and psychological infrastructure to engage with AI and exercise its unique intelligence to benefit humans as much as possible. The technology will compel adaptation in many—indeed, most—aspects of political and social life.

In each discrete major new deployment of AI, it will be crucial to establish the balance. Societies and their leaders will have to choose when individuals should be notified that they are dealing with AI as well as what powers they have in those interactions. Ultimately, through these choices, a new human identity for the AI age will be made manifest.

Some societies and institutions may adapt by degrees. Others, however, may find their foundational assumptions in conflict with the way they have come to perceive reality and themselves. Since AI facilitates education and access to information even as it increases the potential for amplification and manipulability, these conflicts may grow. Better informed, better equipped, and with their viewpoints amplified, individuals may demand more of their governments.

Several principles emerge. First, to ensure human autonomy, core governmental decisions should be carved out of AI-imbued structures and limited to human administration and oversight. Principles inherent in our societies provide for peaceful resolutions of disputes. In this process, order and legitimacy are linked: order without legitimacy is mere force.

Ensuring human oversight of, and determinative participation in, the basic elements of government will be essential to sustaining legitimacy. In the administration of justice, for example, providing explanations and moral reasoning are crucial elements of legitimacy, permitting participants to assess a tribunal's fairness and challenge its conclusions if they fail to accord with societally held moral principles. It follows that in the age of AI, whenever such a significant issue is at stake, the deciders will need to be qualified, non-anonymous humans who can offer reasons for the choices made.

Similarly, democracy must retain human qualities. At the most basic level, this will mean protecting the integrity of democratic deliberations and elections. Meaningful deliberation requires more than the opportunity to speak; it also requires the protection of human speech from AI distortion. Free speech needs to be continued for humans but not extended to AI. As we said in chapter 4, AI has the capacity to generate, both in high quality and large volume, misinformation such as deep fakes, which are very difficult to distinguish from real video and audio recordings. Although automated AI speech was created and deployed at people's behest, it will be important to develop understandable distinctions between it and genuine human speech. Though regulation of AI intermediation that prevents the promotion of misinformation and disinformation—deliberately created falsehoods—will be difficult, it will be crucial. In a democracy, speech permits citizens to share relevant information, to participate deliberatively in the democratic process, and to pursue self-realization through the production of fiction, art, and poetry.[4] AI-generated false statements may approximate human speech, but they serve only to drown it out or distort it. Curbing the spread of AI that produces misinformation, therefore, would help preserve the speech that is vital to our deliberative process. Does one classify an AI dialogue between two public figures who never met as misinformation, entertainment, or political inquiry—or does the answer depend on the context or on the participants? Does an individual have the right not to be represented in a simulated reality without his or her permission? If permission is granted, is the synthetic expression any more genuine?

Each society must determine in the first instance the full range of permissible and impermissible uses of AI in various domains. Access to certain powerful AI, such as AGI, will need to be strictly guarded to prevent misuse. Because AGI will likely be so expensive to build that only a few will be, access may be inherently limited. Certain limits may violate a society's concepts of free enterprise and the democratic process. Others, such as the need to restrict the use of AI in the production of biological weapons, should be readily agreed upon but will require international collaboration.

As of this writing, the EU has outlined plans to regulate AI,[5] seeking to balance European values such as privacy and freedom with the need for economic development and support of European-grown AI companies. The regulations chart a course between that of China, where the state is investing heavily in AI, including for surveillance purposes, and that of the United States, where AI R&D has largely been left to the private sector. The EU's goal is to rein in the ways companies and governments use data and AI *and* facilitate the creation and growth of European AI companies. The regulatory framework includes risk assessments of various uses of AI and imposes limits or even bans on government use of certain technologies deemed high risk, such as facial recognition (though facial recognition has beneficial uses, such as finding missing persons and combating human trafficking). There will undoubtedly be extensive debate and modification of the initial concept, but its first form is an example of a society determining the range of limitations on AI that it believes will enable it to advance its way of life and future.

In time, these efforts will be institutionalized. In the

United States, academic groups and advisory bodies are already beginning to examine the relationships between existing processes and structures and the rise of artificial intelligence. These include efforts in academia, such as the MIT initiative to address the future of work,[6] and efforts in government, such as the National Security Commission on Artificial Intelligence.[7] Some societies may forgo analysis altogether. They will fall behind societies that, because they inquire, adapt their institutions in advance, or, as we discuss in the following chapter, establish completely new institutions, thereby reducing dislocations and maximizing the material and intellectual benefits partnership with AI offers. As AI develops, the establishment of such institutions will be crucial.

PERCEPTIONS OF REALITY AND HUMANITY

Reality explored by AI, or with the assistance of AI, may prove to be something other than what humans had imagined. It may have patterns we have never discerned or cannot conceptualize. Its underlying structure, penetrated by AI, may be inexpressible in human language alone. As one of our colleagues has observed of AlphaZero, "Examples like this show that there are ways of knowing that are not available to human consciousness."[8]

To chart the frontiers of contemporary knowledge, we may task AI to probe realms we cannot enter; it may return with patterns or predictions we do not fully grasp. The prognostications of the Gnostic philosophers, of an inner reality beyond ordinary human experience, may prove newly significant. We may find ourselves one step closer to the concept of pure knowledge, less limited by the structure of our minds and the patterns of conventional human thought. Not only will we have to redefine our roles as something other than the sole knower of reality, we will also have to redefine the very reality we thought we were exploring. And even if reality does not mystify us, the emergence of AI may still alter our engagement with it and with one another.

As AI becomes prevalent, some people may regard humankind as more capable than ever of knowing and organizing its surroundings. Others may declare our capabilities less adept than we had believed. Such redefinitions of ourselves, and of the reality we find ourselves in, will transform basic assumptions—and, with them, social, economic, and political arrangements. The medieval world had its *imago dei,* its feudal agrarian patterns, its reverence for the crown, and its orientation toward the soaring heights of the cathedral spire. The age of reason had its *cogito ergo sum* and its quest for new horizons—and, with it, new assertions of agency within both individual and societal notions of destiny. The age of AI has yet to define its organizing principles, its moral concepts, or its sense of aspirations and limitations.

The AI revolution will occur more quickly than most humans expect. Unless we develop new concepts to explain, interpret, and organize its consequent transformations, we will be unprepared to navigate it or its implications. Morally, philosophically, psychologically, practically—in every way—we find ourselves on the precipice of a new epoch. We must draw on our deepest resources—reason, faith, tradition, and technology—to adapt our relationship with reality so it remains human.

CHAPTER 7

AI AND THE FUTURE

THE CHANGES WROUGHT by advances in printing in fifteenth-century Europe offer a historical and philosophical comparison to the challenges of the age of AI. In medieval Europe, knowledge was esteemed but books were rare. Individual authors produced literature or encyclopedic compilations of facts, legends, and religious teachings. But these books were a treasure vouchsafed to a few. Most experience was lived, and most knowledge was transmitted orally.

In 1450, Johannes Gutenberg, a goldsmith in the German city of Mainz, used borrowed money to fund the creation of an experimental printing press. His effort barely succeeded—his business floundered, and his creditors sued—but by 1455, the Gutenberg Bible, Europe's first printed book, appeared. Ultimately, his printing press brought about a revolution that reverberated across every sphere of Western, and eventually global, life. By 1500, an estimated nine million printed books circulated in Europe, with the price of an individual book having plummeted. Not only was the Bible widely distributed in the languages of day-to-day life (rather than Latin), the works of classical authors in the fields of history, literature, grammar, and logic also began to proliferate.[1]

Before the advent of the printed book, medieval Europeans accessed knowledge primarily through community traditions—participating in harvesting and seasonal cycles, with their accumulation of folk wisdom; practicing faith and observing its sacraments at places of worship; joining a guild, learning its techniques, and being admitted to its specialized networks. When new information was acquired or new ideas arose (news from abroad, an innovative farming or mechanical invention, novel theological interpretations), it was transmitted either orally through a community or manually through hand-copied manuscripts.

As printed books became widely available, the relationship between individuals and knowledge changed. New information and ideas could spread quickly, through more varied channels. Individuals could seek out information useful to their specific endeavors and teach it to themselves. By examining source texts, they could probe accepted truths. Those with strong convictions and access to modest resources or a patron could publish their insights and interpretations. Advances in science and mathematics could be transmitted quickly, at continental scale. The exchange of pamphlets became an accepted method of political dispute, intertwined with theological dispute. New ideas spread, often either toppling or fundamentally reshaping established orders, leading to adaptations of religion (the Reformation), revolutions in politics (adjusting the concept of national sovereignty), and new understandings in the sciences (redefining the concept of reality).

Today, a new epoch beckons. In it, once again, technology will transform knowledge, discovery, communication, and individual thought. Artificial intelligence is not human. It does not hope, pray, or feel. Nor does it have awareness or reflective capabilities. It is a human creation, reflecting human-designed processes on human-created machines. Yet in some instances, at awesome scale and speed, it produces results approximating those that have, until now, only been reached through human reason. Sometimes, its results astound. As a result, it may reveal aspects of reality more dramatic than any we have ever contemplated. Individuals and societies that enlist AI as a partner to amplify skills or pursue ideas may be capable of feats—scientific, medical, military, political, and social—that eclipse those of preceding periods. Yet once machines approximating human intelligence are regarded as key to producing better and faster results, reason alone may come to seem archaic. After defining an epoch, the exercise of individual human reason may find its significance altered.

The printing revolution in fifteenth-century Europe produced new ideas and discourse, both disrupting and enriching established ways of life. The AI revolution stands to do something similar: access new information, produce major scientific and economic advances, and in so doing, transform the world. But its impact on discourse will be difficult to determine. By helping humanity navigate the sheer totality of digital information, AI will open unprecedented vistas of knowledge and understanding. Alternatively, its discovery of patterns in masses of data may produce a set of maxims that become accepted as orthodoxy across continental and global network platforms. This, in turn, may diminish humans' capacity for skeptical inquiry that has defined the current epoch. Further, it may channel certain societies and network-platform communities into separate and contradictory branches of reality.

AI may better or—if wrongly deployed—worsen humanity, but the mere fact of its existence challenges and, in some cases, transcends fundamental assumptions. Until now, humans alone developed their understanding of reality, a capacity that defined our place in the world and relationship to it. From this, we elaborated our philosophies, designed our governments and military strategies, and developed our moral precepts. Now AI has revealed that reality may be known in different ways, perhaps in more complex ways, than what has been understood by humans alone. At times, its achievements may be as striking and disorienting as those of the most influential human thinkers in their heydays—producing bolts of insight and challenges to established concepts, all of which demand a reckoning. Even more frequently,

AI will be invisible, embedded in the mundane, subtly shaping our experiences in ways we find intuitively suitable.

We must recognize that AI's achievements, within its defined parameters, sometimes rank beside or even surpass those that human resources enable. We may comfort ourselves by repeating that AI is artificial, that it has not or cannot match our conscious experience of reality. But when we encounter some of AI's achievements—logical feats, technical breakthroughs, strategic insights, and sophisticated management of large, complex systems—it is evident that we are in the presence of another experience of reality by another sophisticated entity.

Accessed by AI, new horizons are opening before us. Previously, the limits of our minds constrained our ability to aggregate and analyze data, filter and process news and conversations, and interact socially in the digital domain. AI permits us to navigate these realms more effectively. It finds information and identifies trends that traditional algorithms could not—or at least not with equal grace and efficiency. In so doing, it not only expands physical reality but also permits expansion and organization of the burgeoning digital world.

Yet, at the same time, AI subtracts. It hastens dynamics that erode human reason as we have come to understand it: social media, which diminishes the space for reflection, and online searching, which decreases the impetus for conceptualization. Pre-AI algorithms were good at delivering "addictive" content to humans. AI is excellent at it. As deep reading and analysis contracts, so, too, do the traditional rewards for undertaking these processes. As the cost of opting out of the digital domain increases, its ability to affect human thought—to convince, to steer, to divert—grows. As a consequence, the individual human's role in reviewing, testing, and making sense of information diminishes. In its place, AI's role expands.

The Romantics asserted that human emotion was a valid and indeed important source of information. A subjective experience, they argued, was itself a form of truth. The postmoderns took the Romantics' logic a step further, questioning the very possibility of discerning an objective reality through the filter of subjective experience. AI will take the question considerably further, but with paradoxical results. It will scan deep patterns and disclose new objective facts—medical diagnoses, early signs of industrial or environmental disasters, looming security threats. Yet in the worlds of media, politics, discourse, and entertainment, AI will reshape information to conform to our preferences—potentially confirming and deepening biases and, in so doing, narrowing access to and agreement upon an objective truth. In the age of AI, then, human reason will find itself both augmented and diminished.

As AI is woven into the fabric of daily existence, expands that existence, and transforms it, humanity will have conflicting impulses. Confronted with technologies beyond the comprehension of the nonexpert, some may be tempted to treat AI's pronouncements as quasi-divine judgments. Such impulses, though misguided, do not lack sense. In a world where an intelligence beyond one's comprehension or control draws conclusions that are useful but alien, is it foolish to defer to its judgments? Spurred by this logic, a re-enchantment of the world may ensue, in which AIs are relied upon for oracular pronouncements to which some humans defer without question. Especially in the case of AGI (artificial general intelligence), individuals may perceive godlike intelligence—a superhuman way of knowing the world and intuiting its structures and possibilities.

But deference would erode the scope and scale of human reason and thus would likely elicit backlash. Just as some opt out of social media, limit screen time for children, and reject genetically modified foods, so, too, will some attempt to opt out of the "AI world" or limit their exposure to AI systems in order to preserve space for their reason. In liberal nations, such choices may be possible, at least at the level of the individual or the family. But they will not be without cost. Declining to use AI will mean not only opting out of conveniences such as automated movie recommendations and driving directions but also leaving behind vast domains of data, network platforms, and progress in fields from health care to finance.

At the civilizational level, forgoing AI will be infeasible. Leaders will have to confront the implications of the technology, for whose application they bear significant responsibility.

The need for an ethic that comprehends and even guides

the AI age is paramount. But it cannot be entrusted to one discipline or field. The computer scientists and business leaders who are developing the technology, the military strategists who seek to deploy it, the political leaders who seek to shape it, and the philosophers and theologians who seek to probe its deeper meanings all see pieces of the picture. All should take part in an exchange of views not shaped by preconceptions.

At every turn, humanity will have three primary options: confining AI, partnering with it, or deferring to it. These choices will define AI's application to specific tasks or domains, reflecting philosophical as well as practical dimensions. For example, in airline and automotive emergencies, should an AI copilot defer to a human? Or the other way around? For each application, humans will have to chart a course; in some cases, the course will evolve, as AI capabilities and human protocols for testing AI's results also evolve. Sometimes deference will be appropriate—if an AI can spot breast cancer in a mammogram earlier and more accurately than a human can, then employing it will save lives. Sometimes partnership will be best, as in self-driving vehicles that function as today's airplane autopilots do. At other times, though—as in military contexts—strict, well-defined, well-understood limitations will be critical.

AI will transform our approach to what we know, how we know, and even what is knowable. The modern era has valued knowledge that human minds obtain through the collection and examination of data and the deduction of insights through observations. In this era, the ideal type of truth has been the singular, verifiable proposition provable through testing. But the AI era will elevate a concept of knowledge that is the result of partnership between humans and machines. Together, we (humans) will create and run (computer) algorithms that will examine more data more quickly, more systematically, and with a different logic than any human mind can. Sometimes, the result will be the revelation of properties of the world that were beyond our conception—until we cooperated with machines.

AI already transcends human perception—in a sense, through chronological compression or "time travel": enabled by algorithms and computing power, it analyzes and learns through processes that would take human minds decades or even centuries to complete. In other respects, time and computing power alone do not describe what AI does.

ARTIFICIAL GENERAL INTELLIGENCE

Are humans and AI approaching the same reality from different standpoints, with complementary strengths? Or do we perceive two different, partially overlapping realities: one that humans can elaborate through reason and another that AI can elaborate through algorithms? If this is the case, then AI perceives things that we do not and cannot—not merely because we do not have the time to reason our way to them, but also because they exist in a realm that our minds cannot conceptualize. The human quest to know the world fully will be transformed—with the haunting recognition that to achieve certain knowledge we may need to entrust AI to acquire it for us and report back. In either case, as AI pursues progressively fuller and broader objectives, it will increasingly appear to humans as a fellow "being" experiencing and knowing the world—a combination of tool, pet, and mind.

This puzzle will only deepen as researchers near or attain AGI. As we wrote in chapter 3, AGI will not be limited to learning and executing specific tasks; rather, by definition, AGI will be able to learn and execute a broad range of tasks, much like those humans perform. Developing AGI will require immense computing power, likely resulting in their being created by only a few well-funded organizations. Like current AI, though AGI may be readily distributable, given its capacities, its applications will need to be restricted. Limitations could be imposed by only allowing approved organizations to operate it. Then the questions will become: who controls AGI? Who grants access to it? Is democracy possible in a world in which a few "genius" machines are operated by a small number of organizations? What, under these circumstances, does partnership with AI look like?

If the advent of AGI occurs, it will be a signal intellectual, scientific, and strategic achievement. But it does not have to occur for AI to herald a revolution in human affairs.

AI's dynamism and capacity for emergent—in other words, unexpected—actions and solutions distinguish it from prior technologies. Unregulated and unmonitored, AIs could diverge from our expectations and, consequently, our intentions. The decision to confine, partner with, or defer to it will not be made by humans alone. In some cases, it will be dictated by AI itself; in others, by auxiliary forces. Humanity may engage in a race to the bottom. As AI automates processes, permits humans to probe vast bodies of data, and organizes and reorganizes the physical and social worlds, advantages may go to those who move first. Competition could compel deployment of AGI without adequate time to assess the risks—or in disregard of them.

An AI ethic is essential. Each individual decision—to constrain, partner, or defer—may or may not have dramatic consequences, but in the aggregate, they will be magnified. They cannot be made in isolation. If humanity is to shape the future, it needs to agree on common principles that guide each choice. Collective action will be hard, and at times impossible, to achieve, but individual actions, with no common ethic to guide them, will only magnify instability.

Those who design, train, and partner with AI will be able to achieve objectives on a scale and level of complexity that, until now, have eluded humanity—new scientific breakthroughs, new economic efficiencies, new forms of security, and new dimensions of social monitoring and control. Those who do not have such agency in the process of expanding AI and its uses may come to feel that they are being watched, studied, and acted upon by something they do not understand

and did not design or choose—a force that operates with an opacity that in many societies is not tolerated of conventional human actors or institutions. The designers and deployers of AI should be prepared to address these concerns—above all, by explaining to non-technologists what AI is doing, as well as what it "knows" and how.

AI's dynamic and emergent qualities generate ambiguity in at least two respects. First, AI may operate as we expect but generate results that we do not foresee. With those results, it may carry humanity to places its creators did not anticipate. Much like the statesmen of 1914 failed to recognize that the old logic of military mobilization, combined with new technology, would pull Europe into war, deploying AI without careful consideration may have grave consequences. These may be localized, such as a self-driving car that makes a life-threatening decision, or momentous, such as a significant military conflict. Second, in some applications, AI may be unpredictable, with its actions coming as complete surprises. Consider AlphaZero, which, in response to the instruction "win at chess," developed a style of play that, in the millennia-long history of the game, humans had never conceived. While humans may carefully specify AI's objectives, as we give it broader latitude, the paths AI takes to accomplish its objectives may come to surprise or even alarm us.

Accordingly, AI's objectives and authorizations need to be designed with care, especially in fields in which its decisions could be lethal. AI should not be treated as automatic. Neither should it be permitted to take irrevocable actions without human supervision, monitoring, or direct control. Created by humans, AI should be overseen by humans. But in our time, one of AI's challenges is that the skills and resources required to create it are not inevitably paired with the philosophical perspective to understand its broader implications. Many of its creators are concerned primarily with the applications they seek to enable and the problems they seek to solve: they may not pause to consider whether the solution might produce a revolution of historic proportions or how their technology may affect various groups of people. The AI age needs its own Descartes, its own Kant, to explain what is being created and what it will mean for humanity.

Reasoned discussion and negotiation involving governments, universities, and private-sector innovators should aim to establish limits on practical actions—like the ones that govern the actions of people and organizations today. AI shares attributes of some regulated products, services, technologies, and entities, but it is distinct from them in vital ways, lacking its own fully defined conceptual and legal framework. For example, AI's evolving and emergent properties pose regulatory challenges: what and how it operates in the world may vary across fields and evolve over time—and not always in predictable ways. The governance of people is guided by an ethic. AI begs for an ethic of its own—one that reflects not only the technology's nature, but also the challenges posed by it.

Frequently, existing principles will not apply. In the age of faith, courts determined guilt during ordeals in which the accused faced trial by combat and God was believed to dictate victory. In the age of reason, humanity assigned guilt according to the precepts of reason, determining culpability and meting out punishment consistent with notions such as causality and intention. But AIs do not operate by human reason, nor do they have human motivation, intent, or self-reflection. Accordingly, their introduction complicates existing principles of justice being applied to humans. When an autonomous system operating on the basis of its own perceptions and decisions acts, does its creator bear responsibility? Or does the fact that the AI acted sever it from its creator, at least in terms of culpability? If AI is enlisted to monitor signs of criminal wrongdoing, or to assist in judgments of innocence and guilt, must the AI be able to "explain" how it reached its conclusions in order for human officials to adopt them?

At what point and in what contexts in the technology's evolution it should be subject to internationally negotiated restrictions is another essential subject of debate. If attempted too early, the technology may be stymied, or there may be incentives to conceal its capabilities; if delayed too long, it may have damaging consequences, particularly in military contexts. The challenge is compounded by the difficulty of designing effective verification regimes for a technology that is ethereal, opaque, and easily distributed. Official negotiators will inevitably be governments. But forums need to be created for technologists, ethicists, the corporations creating and operating AIs, and others beyond these fields.

For societies, the dilemmas AI raises are profound. Much of our social and political life now transpires on network platforms enabled by AI. This is especially the case for democracies, which depend upon these information spaces for the debate and discourse that form public opinion and confer legitimacy. Who or what institutions should define the technology's role? Who should regulate it? What roles should be played by the individuals who use AI? The corporations that produce it? The governments of the societies that deploy it? As part of addressing such questions, we should seek ways to make it auditable—that is, to make its processes and conclusions both checkable and correctable. In turn, formulating corrections will depend upon the elaboration of principles responsive to AI's forms of perception and decision making. Morality, volition, even causality do not map neatly onto a world of autonomous AIs. Versions of such questions arise for most other elements of society, from transportation to finance to medicine.

Consider AI's impact on social media. Through recent innovations, these platforms have rapidly come to host vital aspects of our communal lives. Twitter and Facebook highlighting, limiting, or outright banning content or individuals—all functions that, as we discussed in chapter 4, depend on AI—are testaments to their power. In particular, democratic nations will be increasingly challenged by the use of AI in the unilateral, often opaque promotion or removal of content and concepts. Will it be possible to retain our agency as our social and political lives increasingly shift into domains curated by AI, domains that we can only navigate through reliance upon that curation?

With the use of AIs to navigate masses of information comes the challenge of distortion—of AIs promoting the world humans instinctually prefer. In this domain, our cognitive biases, which AIs can readily magnify, echo. And with those reverberations, with that multiplicity of choice coupled with the power to select and screen, misinformation proliferates. Social media companies do not run news feeds to promote extreme and violent political polarization. But it is self-evident that these services have not resulted in the maximization of enlightened discourse.

AI, FREE INFORMATION, AND INDEPENDENT THOUGHT

What, then, should our relationship with AI be? Should it be cabined, empowered, or a partner in governing these spaces? That the distribution of certain information—and, even more so, deliberate disinformation—can damage, divide, and incite is beyond dispute. Some limits are needed. Yet the alacrity with which harmful information is now decried, combated, and suppressed should also prompt reflection. In a free society, the definitions of *harmful* and *disinformation* should not be the purview of corporations alone. But if they are entrusted to a government panel or agency, that body should operate according to defined public standards and through verifiable processes in order not to be subject to exploitation by those in power. If they are entrusted to an AI algorithm, the objective function, learning, decisions, and actions of that algorithm must be clear and subject to external review and at least some form of human appeal.

Naturally, the answers will vary across societies. Some may emphasize free speech, possibly differently based on their relative understandings of individual expression, and may thus limit AI's role in moderating content. Each society will choose what it values, perhaps resulting in complex relations with operators of transnational network platforms. AI is porous—it learns from humans, even as we design and shape it. Thus not only will each society's choices vary, so, too, will each society's relationship with AI, its perception of AI, and the patterns that its AIs imitate and learn from human teachers. Nevertheless, the quest for facts and truth should not lead societies to experience life through a filter whose contours are undisclosed and untestable. The spontaneous experience of reality, in all its contradiction and complexity, is an important aspect of the human condition—even when it leads to inefficiency or error.

AI AND INTERNATIONAL ORDER

Globally, myriad questions demand answers. How can AI network platforms be regulated without inciting tensions among countries concerned about their security implications? Will such network platforms erode traditional concepts of state sovereignty? Will the resulting changes impose a polarity on the world not known since the collapse of the Soviet Union? Will small nations object? Will efforts to mediate such consequences succeed, or have any hope of success at all?

As AI's capabilities continue to increase, defining humanity's role in partnership with it will be ever more important and complicated. One can contemplate a world in which humans defer to AI to an ever-greater degree over issues of ever-increasing magnitude. In a world in which an opponent successfully deploys AI, could leaders defending against it responsibly decide not to deploy their own, even if they were unsure what evolution that deployment would portend? And if the AI possessed a superior ability to recommended a course of action, could policy makers reasonably refuse, even if the course of action entailed sacrifice of some magnitude? For what human could know whether the sacrifice was essential to victory? And if it was, would the policy maker truly wish to gainsay it? In other words, we may have no choice but to foster AI. But we also have a duty to shape it in a way that is compatible with a human future.

Imperfection is one of the most enduring aspects of human experience, especially of leadership. Often, policy makers are distracted by parochial concerns. Sometimes, they act on the basis of faulty assumptions. Other times, they act out of pure emotion. Still other times, ideology warps their vision. Whatever strategies emerge to structure the human-AI partnership, they must accommodate. If AI displays superhuman capabilities in some areas, their use must be assimilable into imperfect human contexts.

In the security realm, AI-enabled systems will be so responsive that adversaries may attempt to attack before the systems are operational. The result may be an inherently destabilizing situation, comparable to the one created by nuclear weapons. Yet nuclear weapons are situated in an international framework of security and arms-control concepts developed over decades by governments, scientists, strategists, and ethicists, subject to refinement, debate, and negotiation. AI and cyber weapons have no comparable framework. Indeed, governments may be reluctant to acknowledge their existence. Nations—and probably technology companies—need to agree on how they will coexist with weaponized AI.

The diffusion of AI through governments' defense functions will alter international equilibrium and the calculations that have largely sustained it in our era. Nuclear weapons are costly and, because of their size and structure, difficult to conceal. AI, on the other hand, runs on widely available computers. Because of the expertise and computing resources needed to train machine-learning models, creating an AI requires the resources of large companies or nation-states. Because the application of AIs is conducted on relatively small computers, AI will be broadly available, including in ways not intended. Will AI-enabled weapons ultimately be available to anyone with a laptop, a connection to the internet, and an ability to navigate its dark elements? Will governments empower loosely affiliated or unaffiliated actors to use AI to harass their opponents? Will terrorists engineer AI attacks? Will they be able to (falsely) attribute them to states or other actors?

Diplomacy, which used to be conducted in an organized, predictable arena, will have vast ranges of both information and operation. The previously sharp lines drawn by geography and language will continue to dissolve. AI translators will facilitate speech, uninsulated by the tempering effect of the cultural familiarity that comes with linguistic study. AI-enabled network platforms will promote communication across borders. Moreover, hacking and disinformation will continue to distort perception and evaluation. As complexity increases, the formulation of implementable agreements with predictable outcomes will grow more difficult.

The grafting of AI functionality onto cyber weapons deepens this dilemma. Humanity sidestepped the nuclear paradox by sharply distinguishing between conventional forces—deemed reconcilable with traditional strategy—and nuclear weapons, deemed exceptional. Where nuclear weapons applied force bluntly, conventional forces were discriminating. But cyber weapons, which are capable of both discrimination and massive destruction, erase this barrier. As AI is mapped onto them, these weapons become more unpredictable and potentially more destructive. Simultaneously, as they move through networks, these weapons defy attribution. They also defy detection—unlike nuclear weapons, they may be carried on thumb drives—and facilitate diffusion. And in some forms, they can, once deployed, be difficult to control, particularly given AI's dynamic and emergent nature.

This situation challenges the premise of a rules-based world order. Additionally, it gives rise to an imperative: to develop a concept of arms control for AI. In the age of AI, deterrence will not operate from historical precepts; it will not be able to. At the beginning of the nuclear age, the verities developed in discussions between leading professors (who had government experience) at Harvard, MIT, and Caltech led to a conceptual framework for nuclear arms control that, in turn, contributed to a regime (and, in the United States and other countries, agencies to implement it). While the academics' thinking was important, it was conducted separately from the Pentagon's thinking about conventional war—it was an addition, not a modification. But the potential military uses of AI are broader than those of nuclear arms, and the divisions between offense and defense are, at least currently, unclear.

In a world of such complexity and inherent incalculability, where AIs introduce another possible source of misperception and mistake, sooner or later, the great powers that possess high-tech capabilities will have to undertake a permanent dialogue. Such dialogue should be focused on the fundamental: averting catastrophe and, in so doing, surviving.

AI and other emerging technologies (such as quantum computing) seem to be moving humans closer to knowing reality beyond the confines of our own perception. Ultimately, however, we may find that even these technologies have limits. Our problem is that we have not yet grasped their philosophical implications. We are being advanced by them, but automatically rather than consciously. The last time human consciousness was changed significantly—the Enlightenment—the transformation occurred because new technology engendered new philosophical insights, which, in turn, were spread by the technology (in the form of the printing press). In our period, new technology has been developed, but remains in need of a guiding philosophy.

AI is a grand undertaking with profound potential benefits. Humans are developing it, but will we employ it to make our lives better or to make our lives worse? It promises stronger medicines, more efficient and more equitable health care, more sustainable environmental practices, and other advances. Simultaneously, however, it has the capability to distort or, at the very least, compound the complexity of the consumption of information and the identification of truth, leading some people to let their capacities for independent reason and judgment atrophy.

Other countries have made AI a national project. The United States has not yet, as a nation, systematically explored its scope, studied its implications, or begun the process of reconciling with it. The United States must make all these projects national priorities. This process will require people with deep experience in various domains to work together—a process that would greatly benefit from, and perhaps require, the leadership of a small group of respected figures from the highest levels of government, business, and academia.

Such a group or commission should have at least two functions:

1. Nationally, it should ensure that the country remains intellectually and strategically competitive in AI.
2. Both nationally and globally, it should study, and raise awareness of, the cultural implications AI produces.

In addition, the group should be prepared to engage with existing national and subnational groups.

We write in the midst of a great endeavor that encompasses all human civilizations—indeed, the entire human species. Its initiators did not necessarily conceive of it as such; their motivation was to solve problems, not to ponder or reshape the human condition. Technology, strategy, and philosophy need to be brought into some alignment, lest one outstrip the others. What about traditional society should we guard? And what about traditional society should we risk in order to achieve a superior one? How can AI's emergent qualities be integrated into traditional concepts of societal norms and international equilibrium? What other questions should we seek to answer when, for the situation in which we find ourselves, we have no experience or intuition?

Finally, one "meta" question looms: can the need for philosophy be met by humans *assisted* by AIs, which interpret and thus understand the world differently? Is our destiny one in which humans do not completely understand machines, but make peace with them and, in so doing, change the world?

Immanuel Kant opened the preface to his *Critique of Pure Reason* with an observation:

> Human reason has the peculiar fate in one species of its cognitions that it is burdened with questions which it cannot dismiss, since they are given to it as problems by the nature of reason itself, but which it also cannot answer, since they transcend every capacity of human reason.[2]

In the centuries since, humanity has probed deeply into these questions, some of which concern the nature of the mind, reason, and reality itself. And humanity has made great breakthroughs. It has also encountered many of the limitations Kant posited—a realm of questions it cannot answer, of facts it cannot know fully.

The advent of AI, with its capacity to learn and process information in ways that human reason alone cannot, may yield progress on questions that have proven beyond our capacity to answer. But success will produce new questions, some of which we have attempted to articulate in this book. Human intelligence and artificial intelligence are meeting, being applied to pursuits on national, continental, and even global scales. Understanding this transition, and developing a guiding ethic for it, will require commitment and insight from many elements of society: scientists and strategists, statesmen and philosophers, clerics and CEOs. This commitment must be made within nations and among them. Now is the time to define both our partnership with artificial intelligence and the reality that will result.

AFTERWORD

A NEW REALITY

IN THE INTERVENING months since this book's original publication, a number of evolutions described in these pages have continued apace, and new phenomena have emerged. We revisit our theme to briefly trace these recent developments.

From electric grids to supply chains to drug discovery,[1] AI has proved its revolutionary potential. In the coming years, AI will accelerate progress on climate change mitigation, transform farming, and revolutionize medicine.

These changes may be coming sooner than we thought. We postulated that by 2040, artificial intelligence would be perhaps a million times more powerful than it was in 2021, following Moore's law, which predicts a doubling in computer processing power every two years. While increases in the power of AI are harder to quantify than increases in computing power, it appears that their growth is even more rapid.

For example, the power of large language models, neural networks that underlie much of today's natural language processing, is growing even more rapidly, tripling in fewer than two years. Microsoft's Megatron-Turing model,[2] released in late 2021, and Google's PaLM,[3] released in early 2022, each has more than 525 billion parameters compared to 175 billion for OpenAI's GPT-3, which we wrote about in previous chapters and which was released in June of 2020. These models also perform more impressively than GPT-3 on a wide range of language tasks. OpenAI is also working on its next version of GPT, continuing the race.

Models such as DeepMind's RETRO and OpenAI's GLIDE have improved both efficiency and capacity, able to do more with the same number of model parameters but often using more training data than older models.[4] As these models add nodes, layers, and connections, they can recognize and use additional relationships and patterns. Together, these developments suggest acceleration toward machines with increasingly multifaceted and sophisticated language abilities—and, more broadly, an increased capacity to learn.

These systems have the ability to generate and edit language in a variety of mediums. Prompted by a few words, AI can automatically complete an article or story by predicting what a human author would have written next. It can also translate between languages, solve word problems, and generate "chain of thought" explanations. For example, the 540-billion-parameter version of PaLM is able to provide explanations of a joke such as the following: "Did you see that Google just hired an eloquent whale for their TPU team? It showed them how to communicate between different pods." PaLM offers the explanation: "TPUs are a type of computer chip that Google uses for deep learning. A 'pod' is a group of TPUs. A 'pod' is also a group of whales. The joke is that the whale is able to communicate between two groups of whales, but the speaker is pretending that the whale is able to communicate between two groups of TPUs." Beyond human language, these systems can predict completions of partial computer code, provide potential fixes to errors in computer code, and even generate translations from human text to computer code.

The performance of language models continues to become increasingly impressive on a wide range of tasks, but at the same time it remains important to recall the admonition from chapter 1, in which GPT-3 describes itself as "a language model, and not a reasoning machine." Language models encode what is reflected in human text rather than offering a deep understanding of it, although they may sometimes project the appearance of such deep understanding.

In 2022, machine learning continued to broaden its vision, so to speak. As OpenAI's chief scientist predicted at the end of 2020, language models have "start[ed] to become aware of the visual world."[5] Multimodality, the ability of text-trained language models to process and generate audio and visual media, is a burgeoning field of exploration. The most prominent current example is DALL·E 2 from OpenAI, announced in early 2022.[6] DALL·E 2 can create photographic images or professional-quality artwork based on arbitrary text descriptions. For example, it has generated realistic images of "cats playing chess" and "a living room filled with sand, sand on the floor, with a piano in the room."[7] These are original images, although originality does not always mean creativity. Indeed, arguably, much of the creativity comes from people's experimentation with text descriptions and the collaboration between the human imagination and a highly sophisticated partner that translates text into imagery.

These models, and others like them, demonstrate that computers are gradually accreting new skills. Text-only models are becoming text-and-image models. And several models are being combined to yield improvements. These developments confirm our intuition, mentioned in chapter 3, that skill accretion could be the path to progressively more general intelligence—for example, an intelligence that allows a single system to complete both the verbal and visual tasks we have described.

That single system could take the form of an AI digital companion. This companion would act as an assistant: it would provide information in response to questions, summarize long articles and books, and review and comment on social media. It may suggest or complete recurring purchases by linking with online retailers or ordering groceries automatically; it might recommend films or podcasts; it could even create humorous messages and videos to share with friends. Many of these capabilities already exist on separate systems, but the digital companion would bring them together. Like a human assistant, this companion would become even more useful with training.

Continued advances in AI language models, visual models, and other specialized models will result in AI that graduates from acting as our assistants to serving as our interlocutors and collaborators. Within the next several years, we may be able to craft "AI twins" of ourselves—second selves that speak and behave as we do. And those twins will outlive us.

These trends will amplify the power of global network platforms. As AIs become storytellers and co-creators of our videos, online content will become more entertaining and more prolific, but it will also make it increasingly difficult to distinguish human expression from that of machines. Already we are experiencing a deluge of online content, with the very real potential that AI-created material is, or soon will be, drowning out direct human expression. This may call for an AI-age definition of freedom of speech and expression that distinguishes between utterances originated by humans and utterances generated by machines.

Advances in AI for scientific discovery have also continued to accelerate. In the summer of 2021, Deep Mind released AlphaFold2, the successor to AlphaFold, which predicts the 3D structure of proteins from their amino acid sequence (see chapter 6). AlphaFold2 and work done at the Baker Lab at the University of Washington were named by *Science* as the "2021 Breakthrough of the Year,"[8] generating a new DeepMind database of protein structures containing nearly one million proteins as of spring 2022, with plans to grow it to nearly one hundred million (hundreds of times more than the number of structures that have been experimentally determined). These advances are accelerating a wide range of biological research, including in the important area of protein-protein interactions for problems such as vaccines for the SARS-CoV-2 virus.

The increasing sophistication of the role of AI in scientific discovery is leading us to a world in which AI may be making substantive contributions to inventions at a level that if undertaken by people would make them co-inventors. Patent law in the United States and most nations recognizes only human inventors, potentially leading to situations in which there are inventions with no inventors (or perhaps in which people play a de minimis co-inventor role).

National governments have recognized AI's threat to language: Hungary has commissioned its own large language model so that Hungarian does not automatically become obsolete in the digital realm.[9] Governments have also begun to grapple with digital networks' dilution of communal and national identity. Some states, including China and Russia, have been more aggressive in this effort than others. The balance between risks of anarchy and risks of oppression is being struck differently in different places. The hope for global standards to control disinformation and other side effects of global network platforms is dwindling.

Our societies and alliances are also tackling with renewed vigor the problem of regulation in the face of the significant challenges that AI poses to security, conflict, and world order. These are not theoretical challenges. Leaps and bounds in AI's ability to manufacture and manipulate DNA are creating new dangers in chemical and biological warfare.[10] Specifically, they create the possibility of tailoring a bioweapon to populations that share specific genetic traits. Autonomous weapons have been deployed on the battlefield.[11] The Department of Defense trains its "AI agents" at "OpenAI gym."[12] How to ensure that automatic systems properly calibrate for proportionality and guard against mistakes without losing time—massively compressed by automation—is a permanent challenge. Most recently, technology developed by Clearview AI, a company that "scrapes," or collects, online images for facial-recognition purposes, has been used by Ukraine to identify dead comrades and living Russian soldiers.[13] With what accuracy we do not know.

Regulators are making strides toward controlling AI's effects in civilian and military realms: AI-related bills in the United States, Europe, and China have proliferated, although with quite different approaches and potential regulatory benefits and risks. NATO published an AI defense strategy just after the original release of this book. Its principles for responsible use are aspirational but fundamentally correct: AI applications in defense must be carefully constructed, accountable, understandable, unbiased, and limited to their intended uses and functions.[14]

Yet inherent risks remain. As the capabilities of models grow, their full array of uses becomes nearly impossible to predict with certainty. Probing a model's weaknesses needs to be done en masse, at scale—and not only by the developers of the model. And if one comes to dominate—or if a few models come to dominate, as we expect they will—we face the danger of a technological monoculture that risks proliferating any weaknesses, biases, or limitations inherent in a small number of models across a large range of domains. Even without such risks, the sheer reduction in the diversity and variability of outcomes that results from having a small number of models makes the nature of those models crucially important. Imagine replacing the independent decisions of millions or even billions of people by a handful of people—or, in this case, a handful of AI models.

These computers will operate with a distinctly nonhuman intelligence—or, rather, several nonhuman intelligences. As of 2021, for example, the intelligence of the computer has increasingly ventured into mathematics and detected patterns unseen by the human mind.[15] We will lack a real understanding of these intelligences, even as they mimic human behavior. For a time, we will be limited to observing their patterns and seeking to craft laws that effectively regulate their effects. Eventually, though, if we are to partner with these systems, we will have to develop a theory of mind for something very new.

The stakes of the interaction between humans and machines are rising rapidly. All human knowledge has existed as if on a wheel, each segment connecting to the next, from philosophy to science to politics to history to literature to art and back again.[16] How does this apply to the revolutionary changes wrought by AI? In the first phase, artificial intelligence is likely to add to our wheel of knowledge as it is, pressing the frontiers of each subject into new terrain. This will expand humans' capacity and accelerate learning. It will do so largely within the conceptual frameworks we have known for acquisition of knowledge. This is a development to which, given enough time, we can adjust, so long as we do not become overly dependent on the computer. Indeed, delegation of mundane matters to AI may free time for conceptual thinking and further expansion of human knowledge.

But in a likely second phase, computers may add new types of knowledge, reaching levels that humans themselves cannot accomplish. Reliance on knowledge that can *only* be gained through machines moves mankind into a new reality.

The development of this reality would be unlike the advancement of the Enlightenment. Then, each expansion of human knowledge built upon previous achievements of reason and intellect. As we noted in chapter 2, Montesquieu characterized the progress of Enlightenment thought as "a chain in which every idea precedes one idea and follows another." Lacking a similar progression, AI provides answers without human logical notions of how one idea follows another. This is an altogether different mode of thinking and knowledge. It may feel discomfiting to humans, who rely on reason for a sense of meaning and agency.

This new intelligence and its different form of logic will change human perception of reality, as AlphaZero's new intelligence is changing human perception of chess. Such a phenomenon requires its own philosophy, which will have to be developed. If we fail to do so, we may find ourselves dismissing entirely the old wheel of knowledge. Enthralled by machines that appear to be our friends, fearful of blocking their superhuman speed, and incapable of explaining their new conclusions, we may develop a reverence for computers that approaches mysticism. The roles of history, morality, justice, and human judgment in such a world are unclear.

AI will lead human beings to realms that we cannot reach solely by human reason, now or perhaps ever. Its technical achievements in health and economics promise to make the age of AI an age of abundance. While we celebrate that potential, we recognize that a new reality is emerging. As the stakes rise, our response must meet them.

ACKNOWLEDGMENTS

THIS BOOK, LIKE the discussion it seeks to facilitate, has benefited from the contributions of colleagues and friends across many fields and generations.

Meredith Potter researched, drafted, edited, and facilitated the combining of our views into frameworks with dedication, diligence, and a special sense for the intangible.

Schuyler Schouten joined the project halfway and, through exceptional analysis and writing, advanced its arguments, examples, and narrative.

Ben Daus joined the project last, but when he did, his additional research, informed by his historical knowledge, helped bring it to its conclusion.

Bruce Nichols, our editor and publisher, provided wise counsel, shrewd edits, and patience with our continued revisions.

Ida Rothschild edited every chapter with her characteristic precision and insight.

Mustafa Suleyman, Jack Clark, Craig Mundie, and Maithra

Raghu provided indispensable feedback on the entire manuscript, informed by their experiences as innovators, researchers, developers, and educators.

Robert Work and Yll Bajraktari of the National Security Commission on Artificial Intelligence (NSCAI) commented on drafts of the security chapter with their characteristic commitment to the responsible defense of the national interest.

Demis Hassabis, Dario Amodei, James J. Collins, and Regina Barzilay explained their work—and its profound implications—to us.

Eric Lander, Sam Altman, Reid Hoffman, Jonathan Rosenberg, Samantha Power, Jared Cohen, James Manyika, Fareed Zakaria, Jason Bent, and Michelle Ritter provided additional feedback that made the manuscript more accurate and, we hope, more relevant to readers.

Any shortcomings are our own.

NOTES

PREFACE

1. "AI Startups Raised USD734bn in Total Funding in 2020," *Private Equity Wire,* November 19, 2020, https://www.privateequitywire.co.uk/2020/11/19/292458/ai-startups-raised-usd734bn-total-funding-2020.

CHAPTER 1

1. Mike Klein, "Google's AlphaZero Destroys Stockfish in 100-Game Match," Chess.com, December 6, 2017, https://www.chess.com/news/view/google-s-alphazero-destroys-stockfish-in-100-game-match; https://perma.cc/8WGK-HKYZ; Pete, "AlphaZero Crushes Stockfish in New 1,000-Game Match," Chess.com, April 17, 2019, https://www.chess.com/news/view/updated-alphazero-crushes-stockfish-in-new-1-000-game-match.
2. Garry Kasparov. Foreword. *Game Changer: AlphaZero's Groundbreaking Chess Strategies and the Promise of AI* by Matthew Sadler and Natasha Regan, New in Chess, 2019, 10.
3. "Step 1: Discovery and Development," US Food and Drug Administration, January 4, 2018, https://www.fda.gov/patients/drug-development-process/step-1-discovery-and-development.

4. Jo Marchant, "Powerful Antibiotics Discovered Using AI," *Nature,* February 20, 2020, https://www.nature.com/articles/d41586-020-00018-3.
5. Raphaël Millière (@raphamilliere), "I asked GPT-3 to write a response to the philosophical essays written about it..." July 31, 2020, 5:24 a.m., https://twitter.com/raphamilliere/status/1289129723310886912/photo/1; Justin Weinberg, "Update: Some Replies by GPT-3," *Daily Nous,* July 30, 2020, https://dailynous.com/2020/07/30/philosophers-gpt-3/#gpt3replies.
6. Richard Evans and Jim Gao, "DeepMind AI Reduces Google Data Centre Cooling Bill by 40%," DeepMind blog, July 20, 2016, https://deepmind.com/blog/article/deepmind-ai-reduces-google-data-centre-cooling-bill-40.
7. Will Roper, "AI Just Controlled a Military Plane for the First Time Ever," *Popular Mechanics,* December 16, 2020, https://www.popularmechanics.com/military/aviation/a34978872/artificial-intelligence-controls-u2-spy-plane-air-force-exclusive.

CHAPTER 2

1. Edward Gibbon, *The Decline and Fall of the Roman Empire* (New York: Everyman's Library, 1993), 1:35.
2. This effort was only considered shocking in the West. For millennia, other civilizations' traditions of governance and statecraft have conducted comparable inquiries into national interests and the methods of their pursuit—China's *Art of War* dates back to the fifth century BCE, and India's *Arthashastra* seems to have been composed roughly contemporaneously.
3. The early twentieth-century German philosopher Oswald Spengler identified this aspect of the Western experience of reality as the "Faustian" society, defined by its impulse toward movement into expansive vistas of space and quest for unlimited knowledge. As the title of his major work, *The Decline of the West,* indicates, Spengler held that this cultural impulse, as all others, had its limits—in this case, defined by the cycles of history.

4. Ernst Cassirer, *The Philosophy of the Enlightenment,* trans. Fritz C. A. Koelln and James P. Pettegrove (Princeton, NJ: Princeton University Press, 1951), 14.
5. Eastern traditions reached similar insights earlier, by different routes. Buddhism, Hinduism, and Taoism all held that human beings' experiences of reality were subjective and relative and thus that reality was not simply what appeared before humans' eyes.
6. Baruch Spinoza, *Ethics,* trans. R. H. M. Elwes, book V, prop. XXXI–XXXIII, https://www.gutenberg.org/files/3800/3800-h/3800-h.htm#chap05.
7. The vicissitudes of history have since transformed Königsberg into the Russian city of Kaliningrad.
8. Immanuel Kant, *Critique of Pure Reason,* trans. Paul Guyer and Allen W. Wood, Cambridge Edition of the Works of Immanuel Kant (Cambridge, UK: Cambridge University Press, 1998), 101.
9. See Paul Guyer and Allen W. Wood, introduction to Kant, *Critique of Pure Reason,* 12.
10. Kant coyly placed the divine beyond the realm of human theoretical reason, preserving it for "belief."
11. See Charles Hill, *Grand Strategies: Literature, Statecraft, and World Order* (New Haven, CT: Yale University Press, 2011), 177–185
12. Immanuel Kant, "Perpetual Peace: A Philosophical Sketch," in *Political Writings,* ed. Hans Reiss, trans. H. B. Nisbet, 2nd, enlarged ed., Cambridge Texts in the History of Political Thought (Cambridge, UK: Cambridge University Press, 1991), 114–115.
13. Michael Guillen, *Five Equations That Changed the World: The Power and the Poetry of Mathematics* (New York: Hyperion, 1995), 231–254.
14. Werner Heisenberg, "Ueber den anschaulichen Inhalt der quantentheoretischen Kinematik and Mechanik," *Zeitschrift für Physik,* as quoted in the *Stanford Encyclopedia of Philosophy,* "The Uncertainty Principle," https://plato.stanford.edu/entries/qt-uncertainty/.
15. Ludwig Wittgenstein, *Philosophical Investigations,* trans. G. E. M. Anscombe (Oxford, UK: Basil Blackwell, 1958), 32–34.

16. See Eric Schmidt and Jared Cohen, *The New Digital Age: Reshaping the Future of People, Nations, and Business* (New York: Alfred A. Knopf, 2013).

CHAPTER 3

1. Alan Turing, "Computing Machinery and Intelligence," *Mind* 59, no. 236 (October 1950), 433–460, reprinted in B. Jack Copeland, ed., *The Essential Turing: Seminal Writings in Computing, Logic, Philosophy, Artificial Intelligence, and Artificial Life Plus the Secrets of Enigma* (Oxford, UK: Oxford University Press, 2004), 441–464.
2. Specifically, a Monte Carlo tree search of future moves enabled or foreclosed.
3. James Vincent, "Google 'Fixed' Its Racist Algorithm by Removing Gorillas from Its Image-Labeling Tech," *The Verge*, January 12, 2018, https://www.theverge.com/2018/1/12/16882408/google-racist-gorillas-photo-recognition-algorithm-ai.
4. James Vincent, "Google's AI Thinks This Turtle Looks Like a Gun, Which Is a Problem," *The Verge*, November 2, 2017, https://www.theverge.com/2017/11/2/16597276/google-ai-image-attacks-adversarial-turtle-rifle-3d-printed.
5. And, to a lesser extent, Europe and Canada.

CHAPTER 4

1. Yet some historical episodes provide instructive parallels. For a survey of interactions between centralized power and networks, see Niall Ferguson, *The Square and the Tower: Networks and Power, from the Freemasons to Facebook* (New York: Penguin Press, 2018).
2. While the term *platform* can be used to mean many different things in the digital realm, we use *network platform* specifically to refer to online services with positive network effects.
3. https://investor.fb.com/investor-news/press-release-details/2021/Facebook-Reports-Fourth-Quarter-and-Full-Year-2020-Results/default.aspx.

4. The statistics on removals are publicly reported quarterly. See https://transparency.facebook.com/community-standards-enforcement.
5. See Cade Metz, "AI Is Transforming Google Search. The Rest of the Web Is Next," in *Wired*, February 4, 2016. Since then, advances in AI for search have continued. Some of the most recent are described on Google's blog, *The Keyword* (see Prabhakar Raghavan, "How AI Is Powering a More Helpful Google," October 15, 2020, https://blog.google/products/search/search-on/), such as spelling correction and the ability to search for specific phrases or passages, videos, and numerical results.
6. Positive network effects can be further understood relative to economies of scale. With economies of scale, a large provider can often have a cost advantage and, when that leads to lower pricing, can benefit individual customers or users. But because positive network effects concern a product or service's effectiveness, not just its cost, they are generally considerably stronger than economies of scale.
7. See Kris McGuffie and Alex Newhouse, "The Radicalization Risks Posed by GPT-3 and Advanced Neural Language Models," Middlebury Institute of International Studies at Monterey, Center on Terrorism, Extremism, and Counterterrorism, September 9, 2020, https://www.middlebury.edu/institute/sites/www.middlebury.edu.institute/files/2020-09/gpt3-article.pdf?fbclid=IwARoroLroOYpt5wgr8EOpsIvGL2sEAi5HoPimcGlQcrpKFaG_HDDs3lBgqpU.

CHAPTER 5

1. Carl von Clausewitz, *On War*, ed. and trans. Michael Howard and Peter Paret (Princeton, NJ: Princeton University Press, 1989), 75.
2. This dynamic extends beyond the purely military realm. See Kai-Fu Lee, *AI Superpowers: China, Silicon Valley, and the New World*

Order (Boston and New York: Houghton Mifflin Harcourt, 2018); Michael Kanaan, *T-Minus AI: Humanity's Countdown to Artificial Intelligence and the New Pursuit of Global Power* (Dallas: BenBella Books, 2020).

3. John P. Glennon, ed., *Foreign Relations of the United States,* vol. 19, *National Security Policy, 1955–1957* (Washington, DC: US Government Printing Office, 1990), 61.
4. See Henry A. Kissinger, *Nuclear Weapons and Foreign Policy* (New York: Harper & Brothers, 1957).
5. See, e.g., Department of Defense, "America's Nuclear Triad," https://www.defense.gov/Experience/Americas-Nuclear-Triad/.
6. See, e.g., Defense Intelligence Agency, "Russia Military Power: Building a Military to Support Great Power Aspirations" (unclassified), 2017, 26–27, https://www.dia.mil/Portals/27/Documents/News/Military%20Power%20Publications/Russia%20Military%20Power%20Report%202017.pdf; Anthony M. Barrett, "False Alarms, True Dangers? Current and Future Risks of Inadvertent U.S.-Russian Nuclear War," 2016, https://www.rand.org/content/dam/rand/pubs/perspectives/PE100/PE191/RAND_PE191.pdf; David E. Hoffman, *The Dead Hand: The Untold Story of the Cold War Arms Race and Its Dangerous Legacy* (New York: Doubleday, 2009).
7. For example, the NotPetya malware deployed by Russian operators against Ukrainian financial institutions and government agencies in 2017 eventually spread well beyond targeted entities in Ukraine to power plants, hospitals, shipping and logistics providers, and energy companies in other countries, including Russia itself. As Senator Angus King and Representative Mike Gallagher, chairs of the United States Cyberspace Solarium Commission, said in their March 2020 report, "Like an infection in the bloodstream, the malware spread along global supply chains." See page 8 of *Report of the United States Cyberspace Solarium Commission,* https://drive.google.com/file/d/1ryMCIL_dZ30QyjFqFkkf10MxIXJGT4yv/view.

8. See Andy Greenberg, *Sandworm: A New Era of Cyberwar and the Hunt for the Kremlin's Most Dangerous Hackers* (New York: Doubleday, 2019); Fred Kaplan, *Dark Territory: The Secret History of Cyber War* (New York: Simon & Schuster, 2016).
9. See Richard Clarke and Robert K. Knake, *The Fifth Domain: Defending Our Country, Our Companies, and Ourselves in the Age of Cyber Threats* (New York: Penguin Press, 2019).
10. See., e.g., *Summary: Department of Defense Cyber Strategy 2018*, https://media.defense.gov/2018/Sep/18/2002041658/-1/-1/1/CYBER_STRATEGY_SUMMARY_FINAL.PDF.
11. For illustrative overviews, see Eric Schmidt, Robert Work, et al., *Final Report: National Security Commission on Artificial Intelligence*, March 2021, https://www.nscai.gov/2021-final-report; Christian Brose, *The Kill Chain: Defending America in the Future of High-Tech Warfare* (New York: Hachette Books, 2020); Paul Scharre, *Army of None: Autonomous Weapons and the Future of War* (New York: W. W. Norton, 2018).
12. Will Roper, "AI Just Controlled a Military Plane for the First Time Ever," *Popular Mechanics*, December 16, 2020, https://www.popularmechanics.com/military/aviation/a34978872/artificial-intelligence-controls-u2-spy-plane-air-force-exclusive.
13. See, e.g., "Automatic Target Recognition of Personnel and Vehicles from an Unmanned Aerial System Using Learning Algorithms," SBIR/STTR (Small Business Innovation Research and Small Business Technology Transfer programs), November 29, 2017 ("Objective: Develop a system that can be integrated and deployed in a class 1 or class 2 Unmanned Aerial System...to automatically Detect, Recognize, Classify, Identify...and target personnel and ground platforms or other targets of interest"), https://www.sbir.gov/sbirsearch/detail/1413823; Gordon Cooke, "Magic Bullets: The Future of Artificial Intelligence in Weapons Systems," *Army AL&T*, June 2019, https://www.army.mil/article/223026/magic_bullets_the_future_of_artificial_intelligence_in_weapons_systems.
14. Scharre, *Army of None*, 102–119.

15. See, e.g., United States White House Office, "National Strategy for Critical and Emerging Technologies," October 2020, https://www.hsdl.org/?view&did=845571; Central Committee of the Communist Party of China, *14th Five-Year Plan for Economic and Social Development and 2035 Vision Goals,* March 2021; Xi Jinping, "Strive to Become the World's Major Scientific Center and Innovation Highland," speech to the Academician Conference of the Chinese Academy of Sciences and the Chinese Academy of Engineering, May 28, 2018, in *Qiushi,* March 2021; European Commission, *White Paper on Artificial Intelligence: A European Approach to Excellence and Trust,* March 2020.
16. See, e.g., Department of Defense Directive 3000.09, "Autonomy in Weapon Systems," rev. May 8, 2017, https://www.esd.whs.mil/portals/54/documents/dd/issuances/dodd/300009p.pdf.
17. See, e.g., Schmidt, Work, et al., *Final Report,* 10, 91–101; Department of Defense, "DOD Adopts Ethical Principles for Artificial Intelligence," February 24, 2020, https://www.defense.gov/Newsroom/Releases/Release/Article/2091996/dod-adopts-ethical-principles-for-artificial-intelligence; Defense Innovation Board, "AI Principles: Recommendations on the Ethical Use of Artificial Intelligence by the Department of Defense," https://admin.govexec.com/media/dib_ai_principles_-_supporting_document_-_embargoed_copy_(oct_2019).pdf.
18. See, e.g., Schmidt, Work, et al., *Final Report,* 9, 278–282.
19. Scharre, *Army of None,* 226–228.
20. See, e.g., Congressional Research Service, "Defense Primer: U.S. Policy on Lethal Autonomous Weapon Systems," updated December 1, 2020, https://crsreports.congress.gov/product/pdf/IF/IF11150; Department of Defense Directive 3000.09, § 4(a); Schmidt, Work, et al., *Final Report,* 92–93.
21. Versions of these concepts were initially explored in William J. Perry, Henry A. Kissinger, and Sam Nunn, "Building on George Shultz's Vision of a World Without Nukes," *Wall Street Journal,* May 23, 2021, https://www.wsj.com/articles

/building-on-george-shultzs-vision-of-a-world-without-nukes-11616537900.

CHAPTER 6

1. David Autor, David Mindell, and Elisabeth Reynolds, "The Work of the Future: Building Better Jobs in an Age of Intelligent Machines," MIT Task Force on the Work of the Future, November 17, 2020, https://workofthefuture.mit.edu/research-post/the-work-of-the-future-building-better-jobs-in-an-age-of-intelligent-machines.
2. "AlphaFold: A Solution to a 50-Year-Old Grand Challenge in Biology," DeepMind blog, November 30, 2020, https://deepmind.com/blog/article/alphafold-a-solution-to-a-50-year-old-grand-challenge-in-biology.
3. See Walter Lippmann, *Public Opinion* (New York: Harcourt, Brace and Company, 1922), 11.
4. Robert Post, "Participatory Democracy and Free Speech," *Virginia Law Review* 97, no. 3 (May 2011): 477–478.
5. European Commission, "A European Approach to Artificial Intelligence," https://digital-strategy.ec.europa.eu/en/policies/european-approach-artificial-intelligence.
6. Autor, Mindell, and Reynolds, "The Work of the Future."
7. Eric Schmidt, Robert Work, et al., *Final Report: National Security Commission on Artificial Intelligence,* March 2021, https://www.nscai.gov/2021-final-report.
8. Frank Wilczek, *Fundamentals: Ten Keys to Reality* (New York: Penguin Press, 2021), 205.

CHAPTER 7

1. J. M. Roberts, *History of the World* (New York: Oxford University Press, 1993), 431–432.
2. Immanuel Kant, *Critique of Pure Reason,* trans. Paul Guyer and Allen W. Wood, Cambridge Edition of the Works of Immanuel Kant (Cambridge, UK: Cambridge University Press, 1998), 99.

AFTERWORD: A NEW REALITY

1. Nathan Benaich and Ian Hogarth, *Welcome to State of AI Report 2021*, October 12, 2021, https://www.stateof.ai/2021-report-launch.html.
2. Ali Alvi and Paresh Kharya, "Using DeepSpeed and Megatron to Train Megatron-Turing NLG 530B, the World's Largest and Most Powerful Generative Language Model," *Microsoft Research Blog*, October 11, 2021, https://www.microsoft.com/en-us/research/blog/using-deepspeed-and-megatron-to-train-megatron-turing-nlg-530b-the-worlds-largest-and-most-powerful-generative-language-model/.
3. Sharan Narang and Aakanksha Chowdhery, "Pathways Language Model (PaLM): Scaling to 540 Billion Parameters for Breakthrough Performance," *Google AI Blog*, April 4, 2022, https://ai.googleblog.com/2022/04/pathways-language-model-palm-scaling-to.html.
4. Will Douglas Heaven, "DeepMind Says Its New Language Model Can Beat Others 25 Times Its Size," *MIT Technology Review*, December 8, 2021, https://www.technologyreview.com/2021/12/08/1041557/deepmind-language-model-beat-others-25-times-size-gpt-3-megatron/.
5. Ilya Sutskever, "Fusion of Language and Vision," *The Batch*, December 20, 2020, https://read.deeplearning.ai/the-batch/issue-72/.
6. "Dall·E 2," OpenAI.com, https://openai.com/dall-e-2/.
7. Cade Metz, "Meet Dall-E, the A.I. That Draws Anything at Your Command," *New York Times*, April 6, 2022, https://www.nytimes.com/2022/04/06/technology/openai-images-dall-e.html
8. Robert Service, "Protein Structures for All," *Science*, December 16, 2021, https://www.science.org/content/article/breakthrough-2021.
9. David F. Carr, "Hungarian Gov Teams Up with Eastern European Bank to Develop AI Supercomputer," *VentureBeat*,

December 9, 2021, https://venturebeat.com/2021/12/09/hungarian-gov-teams-up-with-eastern-european-bank-to-develop-ai-supercomputer/.

10. David Gisselsson, "Next-Generation Biowarfare: Small in Scale, Sensational in Nature?," Mary Ann Liebert, Inc., April 22, 2022, https://www.liebertpub.com/doi/full/10.1089/hs.2021.0165.
11. Nathaniel Allen and Marian "Ify" Okpali, "Artificial Intelligence Creeps on to the African Battlefield," *TechStream*, February 2, 2022, https://www.brookings.edu/techstream/artificial-intelligence-creeps-on-to-the-african-battlefield/.
12. Department of Defense Small Business Innovation Research (SBIR) Program: SBIR 22.2 Program Broad Agency Announcement (BAA), April 20, 2022, https://media.defense.gov/2022/Apr/15/2002977654/-1/-1/1/DOD_22.2_FULL.PDF.
13. Paresh Dave and Jeffrey Dastin, "Exclusive: Ukraine Has Started Using Clearview AI's Facial Recognition During War," Reuters, March 14, 2022, https://www.reuters.com/technology/exclusive-ukraine-has-started-using-clearview-ais-facial-recognition-during-war-2022-03-13/.
14. "Summary of the NATO Artificial Intelligence Strategy," NATO, October 22, 2021, https://www.nato.int/cps/en/natohq/official_texts_187617.htm.
15. Kyle Wiggers, "DeepMind Claims AI Has Aided New Discoveries and Insights in Mathematics," *VentureBeat*, December 1, 2021, https://venturebeat.com/2021/12/01/deepmind-claims-ai-has-aided-new-discoveries-and-insights-in-mathematics/.
16. So the late diplomat and professor Charles Hill taught his students.

INDEX

ABOUT THE AUTHORS

HENRY A. KISSINGER served as the 56th Secretary of State from September 1973 until January 1977. He also served as the Assistant to the President for National Security Affairs from January 1969 until November 1975. He received the Nobel Peace Prize in 1973, the Presidential Medal of Freedom in 1977, and the Medal of Liberty in 1986. Presently, he is chairman of Kissinger Associates, an international consulting firm.

ERIC SCHMIDT is a technologist, entrepreneur, and philanthropist. He joined Google in 2001, helping the company grow from a Silicon Valley start-up to a global technological leader. He served as Chief Executive Officer and chairman from 2001 to 2011, and as executive chairman and technical advisor thereafter. Under his leadership, Google dramatically scaled its infrastructure and diversified its product offerings while maintaining a culture of innovation. In 2017, he cofounded Schmidt Futures, a philanthropic initiative that bets early on exceptional people making the world better. He is the host of *Reimagine with Eric Schmidt,* a podcast exploring how society can build a brighter future after the COVID-19 pandemic.

DANIEL HUTTENLOCHER is the inaugural dean of the MIT Schwarzman College of Computing. Previously he helped found Cornell Tech, the digital technology–oriented graduate school created by Cornell University in New York City, and served as its first dean and vice provost. His research and teaching have been recognized by a number of awards, including an ACM fellowship and CASE Professor of the Year. He has a mix of academic and industry background, having been a computer science faculty member at Cornell, a researcher and manager at the Xerox Palo Alto Research Center (PARC), and the CTO of a fintech start-up. He also has served on a number of boards, including the John D. and Catherine T. MacArthur Foundation, Corning, Inc., and Amazon.com. He received his bachelor's degree from the University of Michigan, and master's and doctoral degrees from MIT.